Springer Desktop Editions in Chemistry

L. Brandsma, S.F. Vasilevsky,
H.D. Verkruijsse
Application of Transition Metal Catalysts
in Organic Synthesis
ISBN 3-540-65550-6

H. Driguez, J. Thiem (Eds.)
Glycoscience, Synthesis of Oligosaccharides
and Glycoconjugates
ISBN 3-540-65557-3

H. Driguez, J. Thiem (Eds.)
Glycoscience, Synthesis of Substrate Analogs
and Mimetics
ISBN 3-540-65546-8

K. Faber (Ed.)
Biotransformations
ISBN 3-540-66949-3

W.-D. Fessner (Ed.)
Biocatalysis, From Discovery to Application
ISBN 3-540-66970-1

S. Grabley, R. Thiericke (Eds.)
Drug Discovery from Nature
ISBN 3-540-66947-7

H.A.O. Hill, P.J. Sadler, A.J. Thomson (Eds.)
Metal Sites in Proteins and Models,
Iron Centres
ISBN 3-540- 65552-2

H.A.O. Hill, P.J. Sadler, A.J. Thomson (Eds.)
Metal Sites in Proteins and Models,
Phosphatases, Lewis Acids and Vanadium
ISBN 3-540-65553-0

H.A.O. Hill, P.J. Sadler, A.J. Thomson (Eds.)
Metal Sites in Proteins and Models,
Redox Centres
ISBN 3-540-65556-5

F.J. Leeper, J.C.Vederas (Eds.)
Biosynthesis, Polyketides and Vitamins
ISBN 3-540-66969-8

A. Manz, H. Becker (Eds.)
Microsystem Technology
in Chemistry and Life Sciences
ISBN 3-540-65555-7

P. Metz (Ed.)
Stereoselective Heterocyclic Synthesis
ISBN 3-540-65554-9

H. Pasch, B. Trathnigg
HPLC of Polymers
ISBN 3-540-65551-4

J. Rohr (Ed.)
Bioorganic Chemistry, Deoxysugars,
Polyketides and Related Classes: Synthesis,
Biosynthesis, Enzymes
ISBN 3-540-66971-X

T. Scheper (Ed.)
New Enzymes for Organic Synthesis,
Screening, Supply and Engineering
ISBN 3-540-65549-2

F.P. Schmidtchen (Ed.)
Bioorganic Chemistry,
Models and Applications
ISBN 3-540-66978-7

Springer

Berlin
Heidelberg
New York
Barcelona
Hong Kong
London
Milan
Paris
Singapore
Tokyo

F.J. Leeper, J.C. Vederas (Eds.)

Biosynthesis
Polyketides and Vitamins

 Springer

Dr. Finian J. Leeper
Department of Chemistry
University of Cambridge
Lensfield Road
Cambridge CB2 1EW, Great Britain
E-mail:fjl1@cam.ac.uk

Professor John C. Vederas
Department of Chemistry
University of Alberta
E3-44 Chemistry Building
Edmonton T6G 2G2, Canada
E-mail: John.Vederas@ualberta.ca

Description of the Series

The Springer Desktop Editions in Chemistry is a paberback series that offers selected thematic volumes from Springer chemistry series to graduate students and individual scientists in industry and academia at very affordable prices. Each volume presents an area of high current interest to a broad non-specialist audience, starting at the graduate student level.

Formerly published as hardcover edition in the review series
Topics in Current Chemistry (Vol. 195) ISBN 978-3-540-63418-8

Cataloging-in-Publication Data applied for

ISBN 978-3-540-66969-2
Springer-Verlag Berlin Heidelberg New York

Die Deutsche Bibliothek - CIP-Einheitsaufnahme
Biosynthesis: polyketides and vitamins / F.J. Leeper; J.C. Vederas (eds.) - Berlin; Heidelberg ;
New York; Barcelona; Hong Kong; London; Milan; Paris; Singapore; Tokyo: Springer, 2000
(Springer desktop editions in chemistry)
ISBN 978-3-540-66969-2

Springer-Verlag is a company in the specialist publishing group BertelsmannSpringer
© Springer-Verlag Berlin Heidelberg 2000

The use of general descriptive names, registered names, trademarks, etc. in this publication does not imply, even in the absence of a specific statement, that such names are exempt from the relevant protective laws and regulations and therefore free for general use.

Cover: design & production, Heidelberg
Typesetting: Fotosatz-Service Köhler OHG, Würzburg
Printed on acid-free paper   SPIN: 10720856   02/3020 hu - 5 4 3 2 1 0

Preface

The present volume is the first of two planned to provide state-of-the-art expert reviews of central topics in modern natural products chemistry and secondary metabolism. Many scientists not directly involved in these areas still view the field of natural products as focused primarily on the isolation, structure elucidation, and cataloging of new compounds (i.e. "grind and find"), or on their chemical synthesis. However, two revolutions in experimental techniques since the early 1950's have completely transformed the understanding of chemical and biological relationships between highly diverse natural products.

The first was the use isotopes to label precursors and follow the fate of key atoms during biochemical transformations to a final product. Advances in methodology that began in the mid-1970's, especially using NMR to detect stable isotopes, dramatically expanded the level of mechanistic detail available to study in vivo conversions of secondary metabolites.

The second revolution started in the late 1980's as researchers developed methods to identify, purify and genetically manipulate individual enzymes responsible for the intricate steps leading to complex natural products. This has expanded the field from building an encyclopaedia of natural materials to being able to initiate control of biochemical pathways to secondary metabolites. Recent advances suggest it may soon be possible to rationally manipulate biochemical pathways in vivo to rapidly produce almost any target molecule, including non-natural variants, in substantial quantity. Since a host of important pharmaceutical agents or their precursors are produced by fermentation, the increased understanding and control of biosynthesis will lead not only to improved production of natural products, but also to an arsenal of modified derivatives for combinatorial approaches in drug discovery.

In the current volume, *Tom Simpson* begins with a review of the modern isotopic techniques useful for elucidation of the overall sequence of events in the construction of a secondary metabolite. The examples chosen to illustrate the applications are primarily polyketides, but the techniques can be applied to any class of natural product. In the second chapter, *Jim Staunton and Barrie Wilkinson* explain the individual steps by which non-aromatic polyketides are assembled, and illustrate the genetic and biochemical studies that have recently led to a detailed knowledge of the remarkable machinery that manufactures complex antibiotics like erythromycin from short chain precursors such as propionate. In the third chapter, *Tadhg Begley and co-authors* describe the new genetic and enzymatic investigations that provide our current mechanistic understanding of biosynthesis of more than a dozen vitamins and enzyme cofactors. This review also identifies limitations in our current grasp of steps involved in the formation of these crucially important coenzymes. The final chapter by *Alan Battersby and Finian Leeper* provides a detailed overview of the formation of vitamin B_{12}, one of the "pigments of life" and the most complex natural product to be found in a wide variety of organisms. In this, as in the other chapters, a wealth of fascinating chemistry is uncovered, which leaves one marvelling at the elegance and efficiency of biosynthetic processes.

Finally we would like to point out that the authors were asked to review the developments in their field over the last five to ten years and particularly their own contributions to the field. If any older work or work from other laboratories which has been significant in the development of the field has not been mentioned, it is for this reason only.

Finian J. Leeper, Cambridge   October 1997
John C. Vederas, Edmonton

Contents

Topics in Current Chemistry
Now Also Available Electronically

For all customers with a standing order for **Topics in Current Chemistry** we offer the electronic form via LINK **free of charge**. You will receive a password for free access to the full articles.
Please register at: **http://link.springer.de/series/tcc/reg_form.htm**

If you do not have a standing order you can nevertheless browse through the table of contents of the volumes and the abstracts of each article at:
http://link.springer.de/series/tcc

There you will also find information about the
- Editorial Board
- Aims and Scope
- Instructions for Authors

Application of Isotopic Methods to Secondary Metabolic Pathways

Thomas J. Simpson

School of Chemistry, University of Bristol, Cantock's Close, Bristol, BS8 1TS, Great Britain.
E-mail: tom.simpson@bristol.ac.uk

The application of stable isotope labelling methods to the study of biosynthetic pathways of secondary metabolites is illustrated by describing studies on the biosynthesis of polyketide and terpenoid metabolites of fungal origins. The use of doubly ^{13}C-labelled precursors provides information on the mode of assembly, cyclisation and rearrangements of polyketide and terpenoid carbon skeletons. More detailed information on the oxidation levels of biosynthetic intermediates and oxidative and reductive modifications occurring during later stages of the pathway can be obtained by the use of precursors labelled with ^{13}C in conjunction with ^{2}H and ^{18}O. These methods are illustrated with reference to studies on the mechanistic and stereochemical details of polyketide chain assembly and on compounds of mixed poly-ketide-terpenoid (meroterpenoid) origins. Fluorine labelled precursors can provide information on key steps as well as producing interesting analogues of the natural metabolites. Application of ^{15}N labelling in conjunction with other isotopes, e.g. ^{13}C and ^{18}O, is illustrated by studies on the formation of 3-nitropropanoic acid. Recent work on the enzymes of polyketide biosynthesis also makes use of multiple stable isotope labelling to assist in the elucidation of 3D protein structure and to obtain information on the mode of action of some of these enzymes.

Keywords: Stable isotopes, fluorine, polyketides, fungi, proteins.

Topics in Current Chemistry, Vol. 195
© Springer Verlag Berlin Heidelberg 1998

1
Introduction

The study of biosynthetic pathways received a major impetus in the early 1970s with the advent of pulsed Fourier-transform NMR spectrometers which greatly facilitated the routine determination of ^{13}C NMR spectra of small amounts of natural products. In addition, precursors enriched with ^{13}C and other stable isotopes, e.g. ^2H, ^{18}O, ^{15}N, were becoming more readily available. These were very timely and synergistic developments because structures of natural products were beginning to be determined increasingly by spectroscopic, mainly NMR, methods with little or no recourse to degradative chemistry. In contrast to stable isotopes, classical radioisotope methods necessitated extensive degradative routes to locate precisely the position of incorporation of isotopic label. In addition, molecules of interest were of increasing structural complexity and were often available in very limited amounts. These problems were largely overcome by ^{13}C and other stable isotope labelling methods. The crucial advance was to bring together once again biosynthetic techniques with methods of structure elucidation which complemented each other.

The basic methodology and principles of NMR-based elucidation of isotopic labelling patterns have been extensively reviewed [1–6] and need not be detailed here. However, it is worth emphasising that the use of singly ^{13}C-labelled precursors does not give any information that could not, in principle, be gained by classical radioisotope methods: they merely (!) made easier the determination of complete labelling patterns. The really significant gains come from the use of precursors multiply labelled with ^{13}C or with ^{13}C in conjunction with ^2H, ^{18}O and ^{15}N. The detection of these combinations of isotopic labels by either spin-spin coupling and/or isotopically shifted signals in the NMR spectra of the enriched metabolites allowed the incorporation of whole biosynthetic units to be elucidated for the first time. The biosynthetic unit could be a single bond or, indeed, a multi-atom unit. Thus the integrity of, for example, carbon-carbon, carbon-hydrogen or carbon-oxygen bonds during a complex metabolic pathway could be tested. The course of skeletal and other rearrangements could be traced and, most importantly, the biosynthetic origin of hydrogen, oxygen and nitrogen could be determined and the oxidation levels of the intermediates in the pathway established by indirect methods. This paved the way for the formulation and testing of ideas on, for example, processive pathways of polyketide biosynthesis [7].

This article will describe some of the results which have come from the application of these methods using examples mainly drawn from our studies on polyketide and terpenoid metabolites of fungal origin. These will be extended with examples of work in other laboratories which the author considers to be particularly effective or interesting applications of stable isotope methods. The article will conclude with recent applications of stable isotopes to some of our work on the enzymology of polyketide biosynthesis to emphasise that this new and burgeoning phase of biosynthetic studies brings new and equally effective opportunities for the applications of stable isotope labelling to the study of natural product biosynthesis.

2
Applications of Double ^{13}C Labelling

Amongst the earliest applications of ^{13}C labelling was the use of doubly labelled [1, 2-^{13}C$_2$]acetate to trace the mode of incorporation of intact acetate units into a wide range of metabolites. This has been one of the major recent developments in biosynthetic methodology and permits information to be obtained which would have been impossible or at best extremely difficult to obtain by classical radio-isotope labelling techniques. The basic concept can be illustrated by a model polyketide system (Fig. 1).

If we consider a molecule of acetate in which both carbons are entirely ^{13}C, ([1, 2-^{13}C$_2$]acetate), it contains two adjacent nuclei of spin 1/2 and so they will couple to each other. If this acetate molecule is incorporated intact into a metabolite then, in any individual molecule, those pairs of carbons derived from an originally intact acetate unit must necessarily both be enriched simultaneously and so will show a mutual ^{13}C-^{13}C coupling. Thus if C-1 is enriched, then C-2 must also be enriched. In the resultant ^{13}C NMR spectrum, the natural abundance signal is flanked by ^{13}C-^{13}C coupling satellites (Fig. 1b). By analysing the coupling patterns, information is obtained on the way in which the precursor molecules are assembled on the enzyme surface and on the way the polyketide chain folds up prior to condensation and cyclisation. If at any stage in the biosynthesis the bond between two carbons originally derived from an intact acetate unit is broken, then the ^{13}C-^{13}C coupling is lost and these carbons appear simply as enhanced singlets, as shown for C-3 and C-4 in Fig. 1c. In this way bond cleavage and rearrangement processes occurring during biosynthesis can be detected.

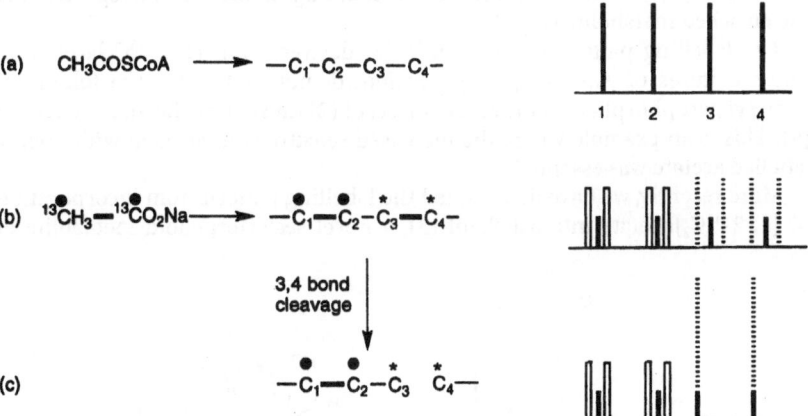

Fig. 1 a–c. Simulated proton noise decoupled ^{13}C NMR spectra of a polyketide-derived moiety: **a** at natural abundance; **b** enriched from [1, 2-^{13}C$_2$]acetate; **c** after cleavage or rearrangement of an originally intact acetate unit

^{13}C-^{13}C Couplings are generally between 30 and 80 Hz and increase in proportion to the amount of "s" character of the atoms in the bond. Hence sp^3-sp^3 bonds are typically 35 Hz; sp^2-sp^3 are 45 Hz, and sp^2-sp^2 are 60 Hz. Substitution on one or both atoms by oxygen increases the size of the coupling. Usually the couplings will be sufficiently different in magnitude to enable the pairs of coupled carbons to be matched up. In addition, if the spectrum has been unambiguously assigned it will be apparent which carbons are mutually coupled.

It should be noted that the labelled precursor is normally highly diluted by the endogenous pool of unlabelled acetate and so no ^{13}C-^{13}C coupling is observed between pairs of atoms derived from different acetate units. In some cases such "extra" inter-acetate couplings may be observed and in a few exceptional cases where little or no dilution by endogenous acetate occurs, the further ^{13}C-^{13}C couplings which result can make interpretation of the spectra problematic. This problem can usually be overcome by the simple expedient of diluting the labelled precursor 2–3-fold with unlabelled material prior to incorporation

The contiguous double labelling technique has been used extensively. This is partly because it extends the permissible dilution factor to at least 2000, as small ^{13}C-^{13}C coupling satellites can be observed with more certainty than the corresponding enrichment from a singly labelled precursor. However, its success is largely because of the extra information obtainable in respect of bond cleavages, rearrangements and symmetrical elements of intermediates. Some applications of the technique are illustrated below.

Incorporation of [1, 2-^{13}C$_2$]acetate is invaluable in distinguishing among alternative foldings of linear polyketide or terpenoid precursors prior to cyclisation and subsequent modifications. Thus, of the three possible foldings of the heptaketide precursor of herqueichrysin (1), a phenalenone metabolite of *Penicillium herquei*, the specific folding shown in Scheme 1 was unambiguously established [8] from the observed ^{13}C-^{13}C couplings in the ^{13}C NMR spectrum of the enriched metabolite (Fig. 2).

The labelling pattern of paniculide (2) derived from [1, 2-^{13}C$_2$]acetate in callus cultures of *Andrographis paniculata* demonstrates that the folding of farnesyl pyrophosphate in the biosynthesis of (2) must be as shown in Scheme 2 [9]. This is an example where the increased sensitivity associated with doubly labelled acetate was essential.

More recently, we have determined the labelling pattern from incorporation of [1, 2-^{13}C$_2$]acetate into astellatol (3), a novel sesterterpenoid metabolite of

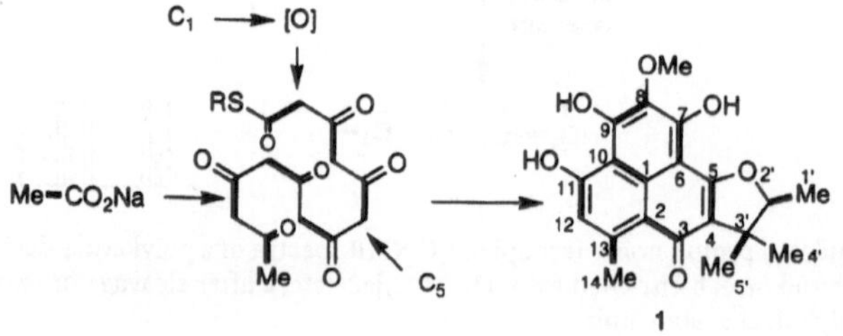

Scheme 1

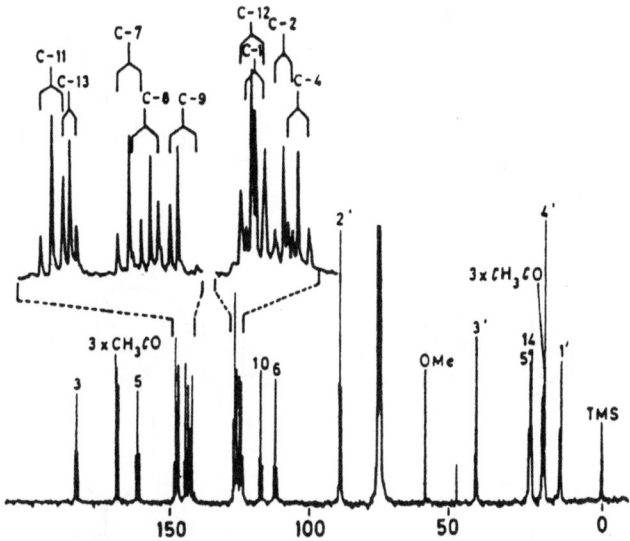

Fig. 2. ¹³C NMR spectrum of [1, 2-¹³C₂]acetate enriched herqueichrysin (as its triacetate) determined at 67.89 MHz. Note the severe overlap of the coupled ¹³C signals in the aromatic region and the second-order ¹³C-¹³C couplings arising from the similarity of the chemical shifts of C-8 and C-9

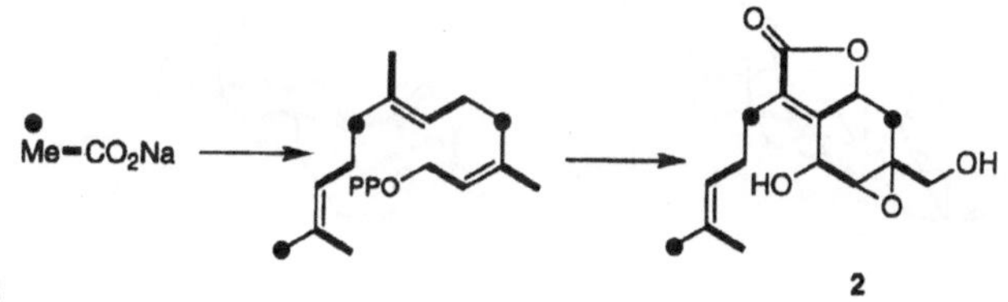

Scheme 2   **2**

Aspergillus stellatus. This supports its biosynthesis via cyclisation of geranyl-farnesyl pyrophosphate followed by rearrangement as shown in Scheme 3 [10]. Surprisingly, the folding pattern must be different from that involved in the bio-synthesis of stellatic acid (**4**), a previously isolated sesterterpene metabolite of *A. stellatus* [11].

The use of ¹³C₂-acetate labelling to detect the involvement of symmetrical intermediates has found extensive use in biosynthetic studies. Thus, as shown in Scheme 4, incorporation of [1, 2-¹³C₂]acetate into ravenelin (**7**) in cultures of *Helminthosporium ravenelii* resulted in the observation of a randomisation of labelling in ring-C which confirmed the predicted involvement of a symmetri-cal benzophenone intermediate (**6**), itself derived from cleavage of an octaketide-derived anthraquinone, presumably helminthosporin (**5**) [12].

The more complex xanthone metabolite, tajixanthone (**11**), was similarly shown to be formed via a symmetrical intermediate. Incorporation of [1,

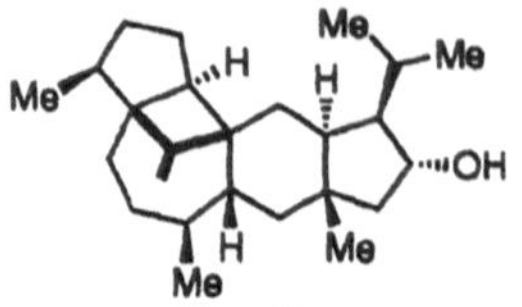

3

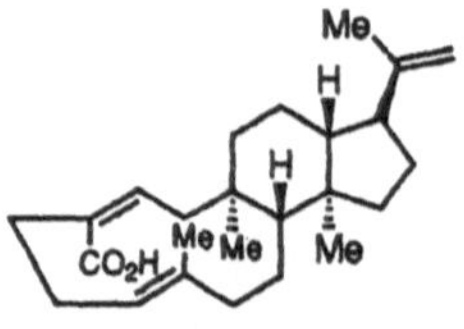

4

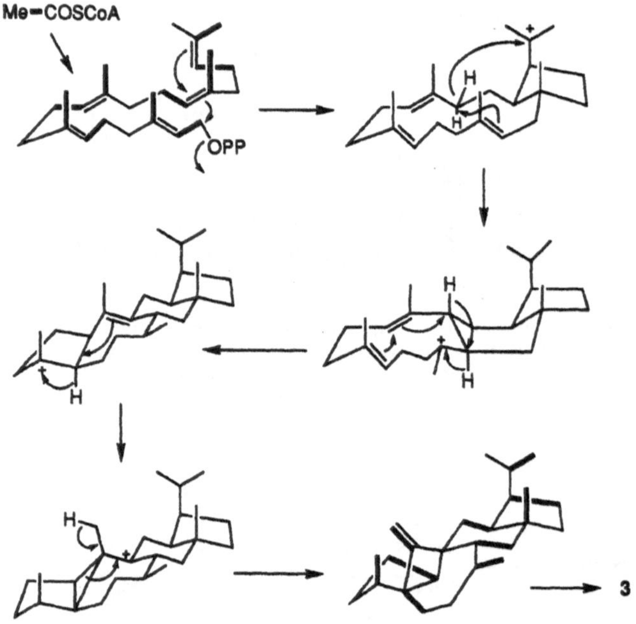

Scheme 3

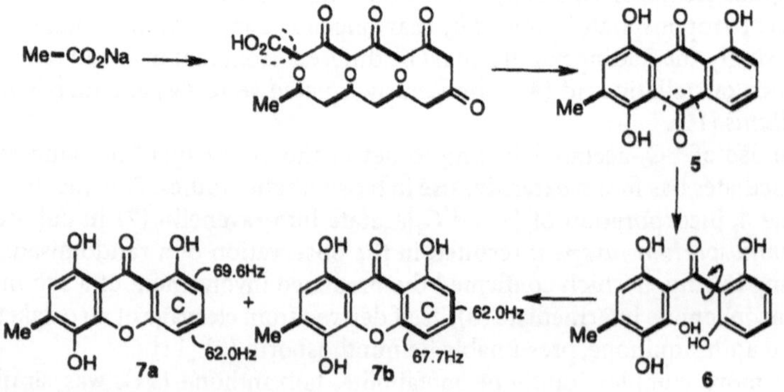

Scheme 4

2-[13C₂]- and [2H₃]acetate gave the results summarised in Scheme 5 [13]. The absence of ^{2}H label on C-25 and C-5 indicated that cleavage of an anthraquinone rather than an anthrone octaketide intermediate occurred and that decarboxylation of the octaketide precursor occurs after cyclisation and aromatisation. The observed scrambling of ^{13}C-^{13}C couplings in ring C implies the involvement of a symmetrical benzophenone intermediate (8) which in turn means that ring cleavage of the anthraquinone precursor must precede introduction of the C-prenyl residue. The specificity of the labelling in the dihydropyran ring, however, suggests that it is formed from an O-prenylaldehyde intermediate (9) by a concerted "ene" reaction.

In vitro studies of the ring closure reaction on closely related xanthone models result in a cis-orientation of the hydroxyl and isopropenyl substituents which can be explained by steric clash between the ketone and xanthone carbonyls in the potential intermediate (10) which can be avoided if the "ene" reaction occurs on the benzophenone intermediate (9). This observation was supported by the isolation of previously known Aspergillus rugulosus metabolites, e.g. arugosin A (12), as co-metabolites of tajixanthone in Aspergillus variecolor. The involvement of anthraquinone intermediates was subsequently

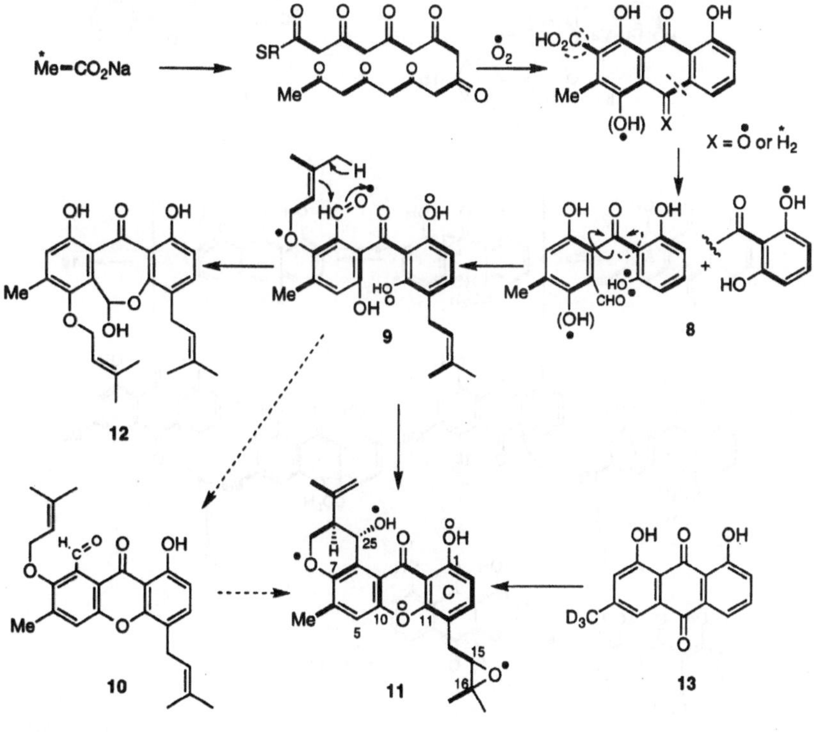

Scheme 5

established when [*methyl-²H₃*]chrysophanol (13) was shown to be a specific precursor for tajixanthone. ²H NMR analysis showed that specific incorporation of label into the aromatic methyl group of (11) had occurred [14]. Further information on the mechanism of xanthone ring closure was obtained by analysis of the ¹³C NMR spectrum and mass spectrum of tajixanthone isolated from *A. variecolor* grown under an atmosphere containing ¹⁸O₂. The intensities of the isotopically shifted signals for C-1, C-10 and C-11 are half those observed elsewhere in the molecule, e.g. C-7, C-25, C-15 and C-16, again consistent with the pathway outlined in Scheme 5.

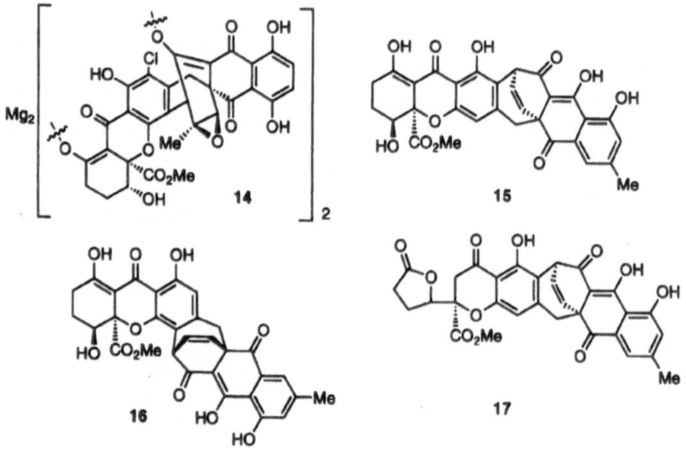

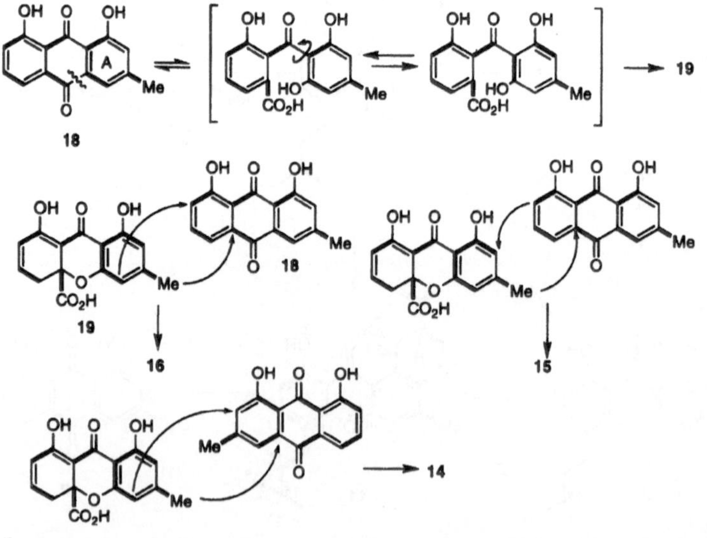

Scheme 6

Further modifications of this anthraquinone ring cleavage pathway can give rise to structurally and biosynthetically more complex metabolites. *Cercospora beticola* toxin which was found to be a 2:2 complex of two magnesium ions with two molecules of (14) [15] and the xanthoquinodins A1 (15) and B1 (16), anti-coccidial metabolites of *Humicola* sp. OF-888 [16] have labelling patterns from incorporation of [$^{13}C_2$]acetate consistent with a heterodimer origin from alternative, and as yet mechanistically obscure, couplings of an anthraquinone (18), and the derived xanthone moiety (19), as indicated in Scheme 6. In the case of these metabolites, the acetate incorporation pattern shows randomisation of labelling in the xanthone ring derived from the A-ring of the anthraquinone (18). In xanthoquinodin A3 (17), it is evident that yet further ring cleavage of the xanthone-derived moiety has occurred.

In marked contrast to ravenelin and tajixanthone, no randomisation of [1, 2-$^{13}C_2$]acetate labelling was observed on incorporation into sterigmatocystin (20) in *Aspergillus variecolor*, ruling out the involvement of a proposed symmetrical benzophenone intermediate [17].

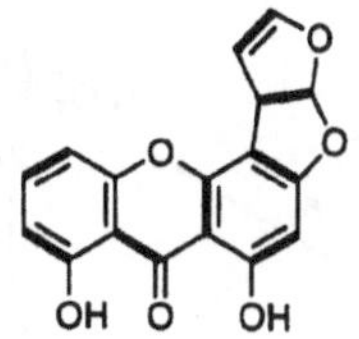

20

A further illustration of the use of ^{13}C-labelled acetates to detect symmetrical intermediates is provided by LL-D253α, a chromanone metabolite of a number of *Phoma* species. Incorporation of [1, 2-$^{13}C_2$]acetate into LL-D253α (21) in *Phoma pigmentivora* indicated that the molecule was assembled from intact acetate units as shown in Scheme 7 [18]. Thus the molecule must be formed from the

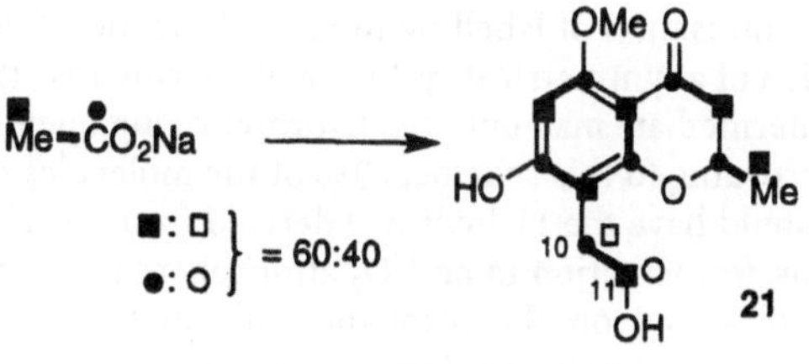

Scheme 7

condensation of two separate preformed polyketide precursors. A particularly surprising feature of this study was the observation of **partial** randomisation of label from incorporation of singly ^{13}C-labelled acetates between C-10 and C-11 in the hydroxyethyl side chain. This is an unusual example where singly ^{13}C-labelled acetate provided more revealing information than [1, 2-$^{13}C_2$]acetate!

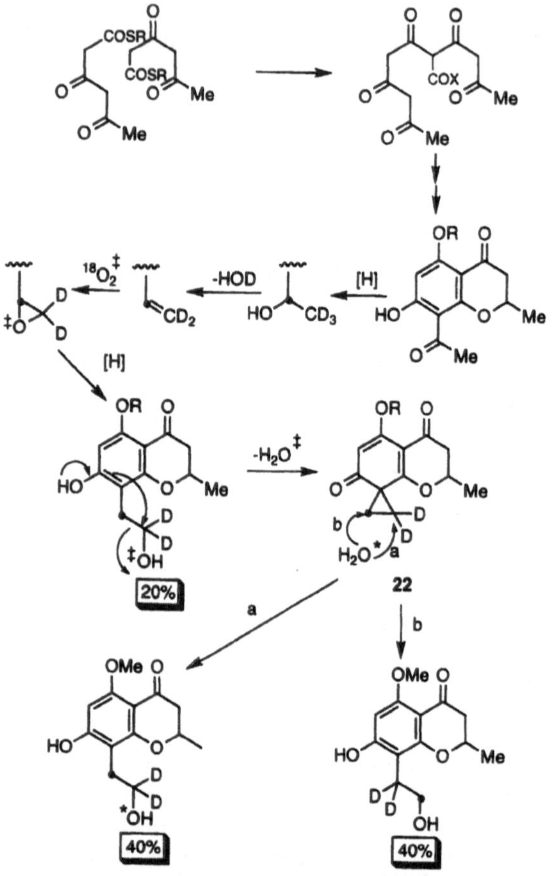

Scheme 8

The observed randomisation of labelling in 80% of the molecules is accounted for by the formation of a symmetrical cyclopropyl intermediate (**22**) as shown in Scheme 8. This intermediate may undergo hydrolytic ring opening at either the α or β carbon. According to this scheme, 20% of the molecules **not** undergoing randomisation should have the 11-hydroxyl derived from the atmosphere and, in accord with this, fermentation in an $^{18}O_2$ atmosphere resulted in an ^{18}O isotope shift being observed on the resonance due to C-11 in the ^{13}C NMR spectrum, the intensity of the shifted signal being approximately 20% that of the unshifted signal. The mass spectrum showed an M+2 peak which mass matched for an ion containing one ^{18}O atom which was approximately 20% of the molecular ion. It is not clear whether the randomisation occurs in vivo or in vitro, but it is noteworthy that an exactly analogous randomisation was recently observed for phomalone (**23**), a butyrophenone metabolite of the fungus *Phoma exigua* [19].

An essential, and often overlooked, feature of these studies is the necessity to have a completely unambiguous assignment of the ^{13}C NMR and ^{1}H NMR spectra of these molecules and this is often the major and most demanding part of the work. An interesting facet of the work on LL-D253α is that the initial assignment studies of the ^{13}C NMR spectrum revealed that the structure of the molecule which had been assigned independently as (24) by three different groups was in fact wrong! Detailed analysis of the long-range ^{1}H-^{13}C couplings in the ^{13}C NMR spectrum led to the revision of the structure which was subsequently proved by total synthesis of both the correct and the previously assigned structures [20].

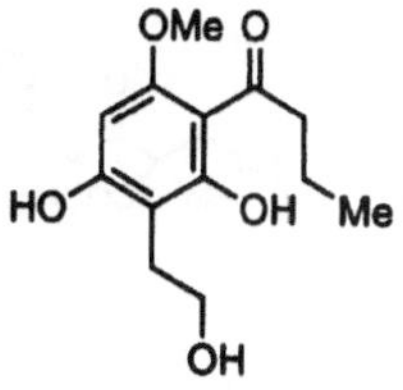

23 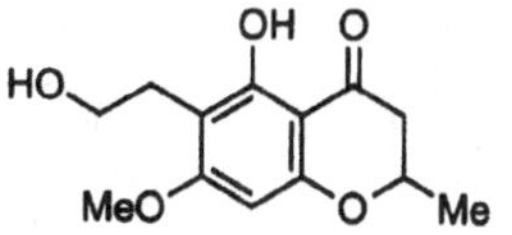   **24**

Incorporations of [^{13}C$_2$]acetate have also enabled the detection of bond cleavage and skeletal rearrangement processes occurring during the biosynthesis of a wide range of metabolites. One of the earliest examples was in the biosynthesis of multicolic acid (26) and related tetronic acid metabolites of *Penicillium multicolor* [21]. The observation of ^{13}C–^{13}C couplings and, more significantly, their absence on the ^{13}C resonances of certain carbons led to the proposal that the tetronic acids were biosynthesised via oxidative cleavage of an aromatic intermediate as shown in Scheme 9. These proposals were subsequently confirmed [22] by the incorporation of 6-pentyl-resorcylic acid (25) and from ^{18}O-labelling studies [23].

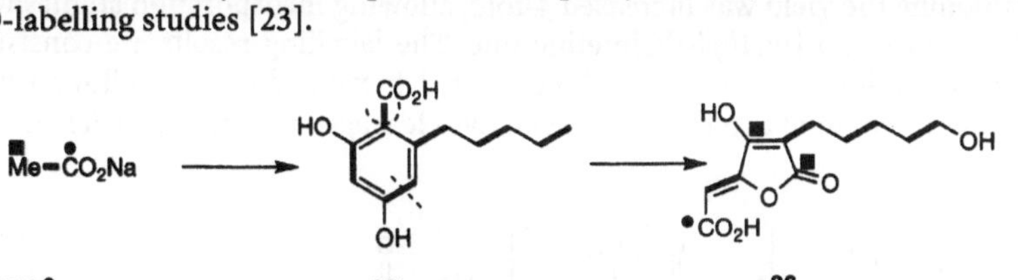

Scheme 9   **25**   **26**

Similar labelling studies on aspyrone (27), a metabolite of *Aspergillus melleus*, gave the labelling pattern shown in Scheme 10 [24] which suggested that its bio-

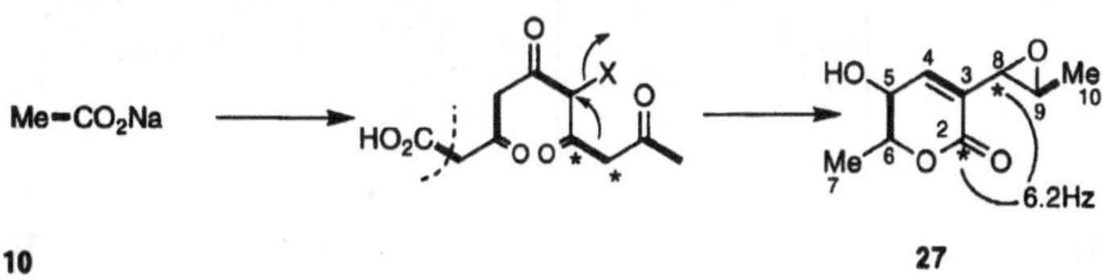

Scheme 10   **27**

synthesis was either via a ring cleavage pathway or by rearrangement. The rearrangement pathway shown was supported by the observation of a two-bond ^{13}C-^{13}C coupling of 6.2 Hz between C-2 and C-8 in the ^{13}C NMR spectrum of [$^{13}C_2$]acetate-enriched aspyrone. This was the first observation of such a coupling in biosynthetic studies. Further work which has led to a complete delineation of the biosynthesis of aspyrone is discussed below.

An extreme example of diverse structures which can arise by extensive and varied modifications of a single polyketide-derived metabolite is provided by the phytotoxic spiciferones (**28**)–(**30**), spiciferinone (**31**) and spiciferin (**32**) produced by the fungus *Cochliobolus spicifer* responsible for leaf spot disease in

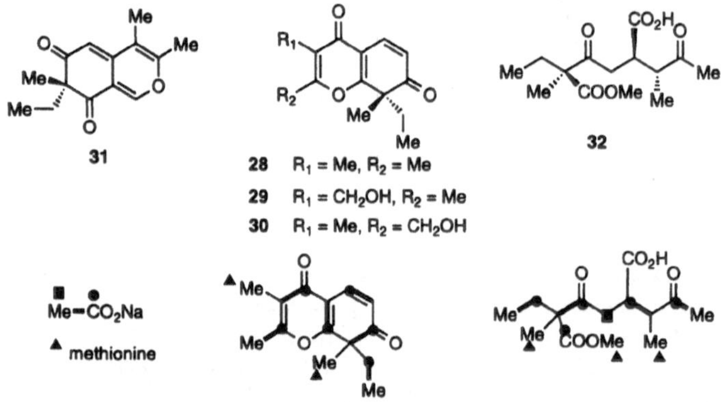
31
28 R1 = Me, R2 = Me
29 R1 = CH2OH, R2 = Me
30 R1 = Me, R2 = CH2OH
32
Me–CO2Na
methionine

wheat. Despite their markedly different carbon skeletons, they were proposed to arise from a common polyketide precursor [25]. Preliminary biosynthetic experiments were hindered by the low yields of the metabolites, but on addition of methionine the yield was increased 4-fold, allowing incorporation studies with [$^{13}C_2$]acetate and [*methyl*-^{13}C]methionine. The labelling results are consistent with cyclisation of a single di-*C*-methylated hexaketide chain followed by a series of retro-aldol reactions and oxidative cleavages as outlined in Scheme 11.

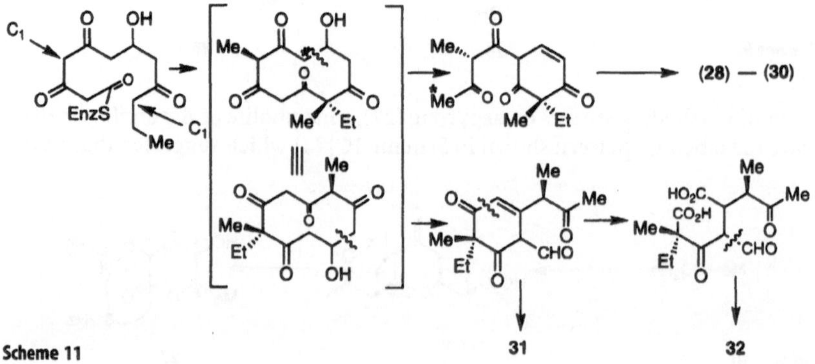
EnzS
(28) — (30)
31
32

Scheme 11

3
Assembly of Highly Reduced Polyketide Metabolites: ^{13}C, ^2H and ^{18}O-Labelling

Despite their enormous structural diversity, polyketide metabolites are related by their common derivation from highly functionalised carbon chains whose assemblies are controlled by multifunctional enzyme complexes, the polyketide synthases (PKSs) which, like the closely related fatty acid synthases, catalyse repetitious sequences of decarboxylative condensation reactions between simple acyl thioesters and malonate, as shown in Fig. 3 [7]. Each condensation is followed by a cycle of modifying reactions: ketoreduction, dehydration and enoyl reduction. In contrast to fatty acid biosynthesis where the full cycle of essentially reductive modifications normally follow each condensation reduction, the PKSs can use this sequence in a highly selective and controlled manner to assemble polyketide intermediates with an enormous number of permutations of functionality along the chain. As shown in Fig. 3, the reduction sequence can be largely or entirely omitted to produce the classical polyketide intermediate which bears a carbonyl on every alternate carbon and which normally cyclises to aromatic polyketide metabolites. On the other hand, the reductive sequence can be used fully or partially after each condensation to produce highly functionalised intermediates such as the "Reduced polyketide" in Fig. 3. Basic questions to be answered are (i) what is the actual polyketide intermediate

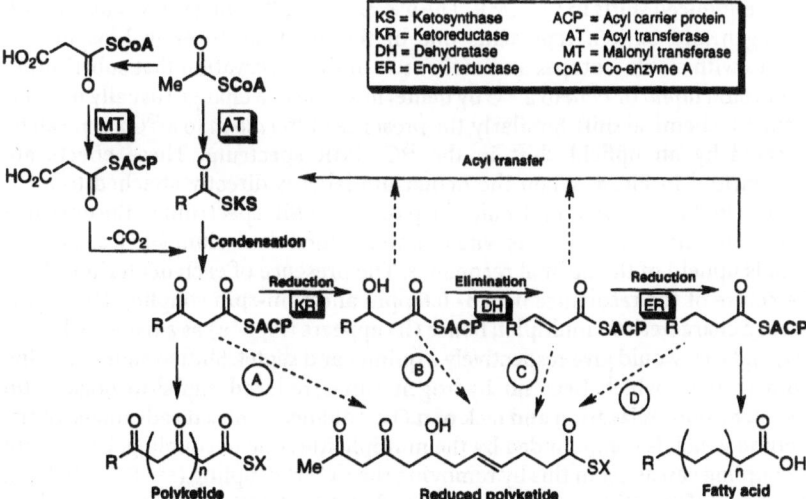

Fig. 3. The assembly of fatty acids, polyketides and reduced polyketides. The "reduced polyketide" intermediate would be formed from an acetate starter by five successive condensation cycles. The first two cycles are condensations and are followed by condensation-ketoreduction, condensation-ketoreduction-elimination, and finally a full condensation-ketoreduction-elimination-enoyl reduction cycle. Thus the overall reaction sequence is A, A, AB, ABC, ABCD

assembled by the PKS; (ii) what is the exact sequence of condensation and reduction steps catalysed by the PKS; and (iii) what further oxidative, reductive and other modifications are required to convert the product of the PKS-catalysed assembly sequence to the final observed metabolite?

Much of our current understanding of the polyketide metabolites formed via these highly reduced intermediates has come from studies using ^{2}H and ^{18}O labelled precursors in conjunction with detailed ^{13}C and ^{2}H NMR analysis of the enriched metabolites [5, 6].

^{2}H NMR, despite several inherent disadvantages, has been the nucleus of choice in many biosynthetic studies. Its major limitations are mainly as a consequence of the low magnetogyric ratio and the relaxation behaviour of the ^{2}H nucleus. Because it is a quadrupole nucleus (spin 1) and thus very efficiently relaxed, the spectral lines are rather broad and this, coupled with the low magnetogyric constant and the small chemical shift range for hydrogen nuclei, often results in poorly resolved spectra. However, the rapid relaxation and lack of any n.O.e. mean that accurate integration of ^{2}H NMR spectra is possible so that the relative enrichment at different sites in a metabolite can be accurately assessed. Another major advantage is that as a consequence of its low natural abundance (0.012 %), much greater dilutions are tolerable than in the case of ^{13}C-labelling: a 100 % ^{2}H-labelled precursor may be diluted 6000-fold and still result in a doubling of intensity over the corresponding natural abundance signal. This makes ^{2}H-labelling particularly suitable for studying the incorporation of advanced intermediates on a biosynthetic pathway [2, 3].

The inherent lack of resolution in ^{2}H NMR can be overcome by the use of isotope-induced shifts in ^{13}C NMR. The use of ^{13}C as a "reporter" nucleus for both hydrogen and oxygen represents one of the great advances in biosynthetic studies with stable isotopes and makes use of the observation that substitution of a proton *alpha* or *beta* to a ^{13}C by deuterium causes a change (usually upfield) in the ^{13}C chemical shift. Similarly, the presence of ^{18}O *alpha* to a ^{13}C atom can be detected by an upfield shift in the ^{13}C NMR spectrum. These effects are summarised in Fig. 4. When the deuterium label is directly attached to a ^{13}C nucleus in the precursor molecule, the p.n.d. ^{13}C NMR spectrum of the enriched metabolite shows, for carbons which have retained deuterium label, a series of signals upfield of the normal resonance. The presence of each deuterium shifts the centre of the resonance by 0.3–0.6 ppm and spin-spin coupling ($^1J_{CD}$) produces a characteristic multiplet; hence CD appears (Fig. 4a) as a triplet, whereas CD$_2$ and CD$_3$ would give respectively a quintet and septet. Shifted signals arising from carbons which bear no hydrogen suffer reduced signal-to-noise ratio caused by poor relaxation and lack of n.O.e. enhancement, a disadvantage of the method which is compounded by the multiplicities due to coupling. Deuterium decoupling can assist in this by removing the ^{13}C-^{2}H coupling (see Fig. 7 below). However, information not obtainable by direct ^{2}H NMR spectroscopy, such as the distribution of label as CH$_2$D, CHD$_2$ and CD$_3$ and the integrity of carbon-hydrogen bonds during biosynthesis, may be gained.

Many of the problems associated with directly attached deuterium are avoided by placing the deuterium label two bonds away from the ^{13}C reporter nucleus. The isotope shift, although reduced, is still observable, and as β-hy-

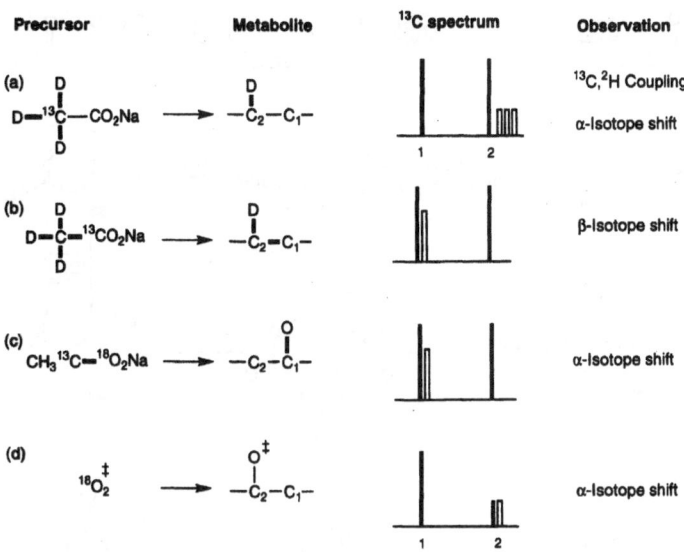

Precursor	Metabolite	^{13}C spectrum	Observation

Fig. 4a–d. Simulated proton noise decoupled ^{13}C NMR spectra of a polyketide-derived moiety from: **a** $[2\text{-}^{13}C, 2\text{-}^{2}H_3]$acetate; **b** $[1\text{-}^{13}C, 2\text{-}^{2}H_3]$acetate; **c** $[1\text{-}^{13}C, ^{18}O_2]$acetate; **d** $^{18}O_2$ gas

drogens only contribute markedly to the relaxation of non-protonated ^{13}C nuclei, the shifted signals otherwise retain any n.O.e. also experienced by the unshifted signals on proton decoupling. As geminal carbon-proton coupling constants are generally small, and carbon-deuterium couplings are over six times smaller again, the shifted signals are effectively singlets (Fig. 4b), even without deuterium decoupling, and this gives a further increase in the signal-to-noise ratio compared with the corresponding α-shift experiment. However, neither of these methods provides reliable information on the stereospecificity of deuterium labelling. Although ^{2}H NMR spectra are disadvantaged by their inherently low dispersion and broad lines, they have the advantage of providing information on the stereospecificity as well as regiospecificity of labelling. ^{2}H NMR, however, does not prove the number of deuteriums incorporated.

The biosynthetic incorporation of ^{18}O can also be detected by the observation of ^{18}O isotope induced shifts in the ^{13}C NMR spectrum, as shown in Fig. 4c, d. The ^{18}O may be conveniently introduced via a doubly labelled precursor or by growth in an $^{18}O_2$ atmosphere. The resulting shifts are generally not much larger than 0.05 ppm. These are very small effects, the same general size as β-^{2}H isotope shifts and are only readily observed with high field spectrometers. These techniques for elucidation of the origins of hydrogen and oxygen provided the basis for much of the work described below.

Colletodiol (**38**), a macrodiolide containing a 14-membered ring, was originally isolated from the plant pathogen *Colletotrichium capsici* along with a number of related macrodiolides including colletoketol (**39**), which was subsequently isolated (as grahamimycin A) from culture filtrates of *Cytospora* sp.

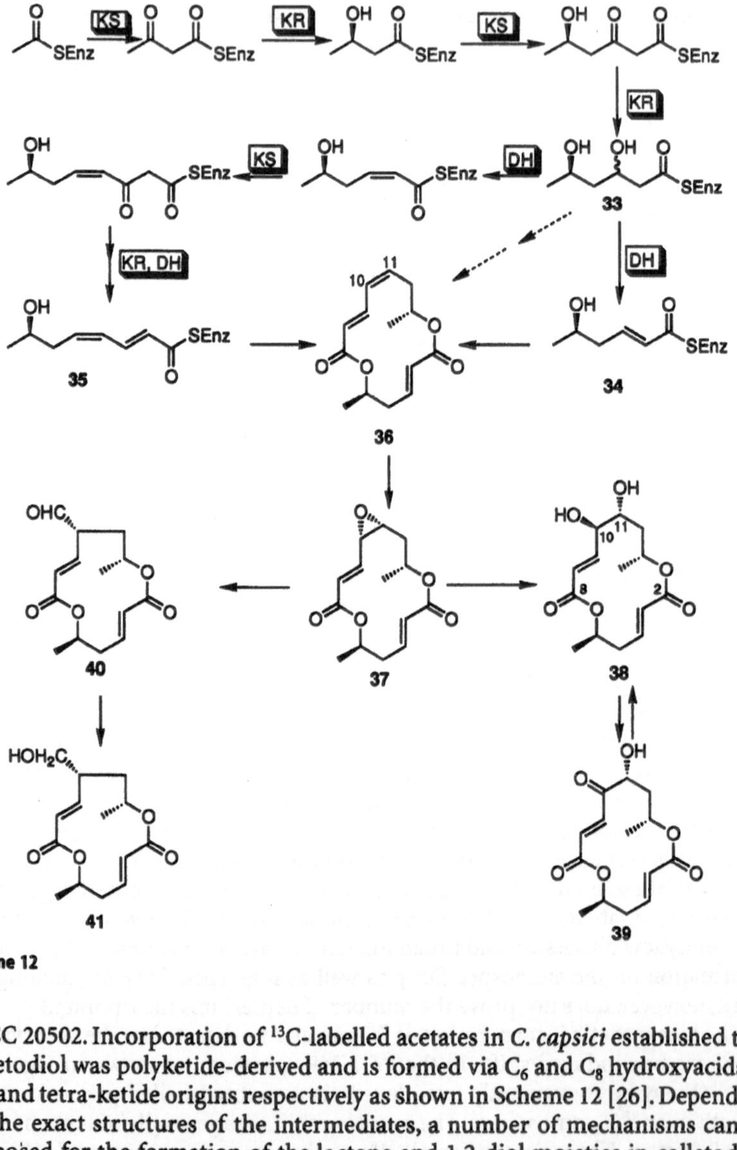

Scheme 12

ATCC 20502. Incorporation of ^{13}C-labelled acetates in *C. capsici* established that colletodiol was polyketide-derived and is formed via C_6 and C_8 hydroxyacids of tri- and tetra-ketide origins respectively as shown in Scheme 12 [26]. Depending on the exact structures of the intermediates, a number of mechanisms can be proposed for the formation of the lactone and 1,2-diol moieties in colletodiol. The origins of all the oxygen and hydrogen atoms have been elucidated by incorporation of [1-13C,2H3]- and [1-13C,18O2]acetates and 18O2 gas by cultures of *Cytospora* [27]. The labelling pattern in colletodiol is summarised in Fig. 5. Interestingly, no ^{2}H isotope-induced shifts could be observed for C-2 or C-8 in the ^{13}C NMR spectrum of the [1-13C,2H3]acetate-enriched colletodiol. It is known [28, 29] that carbonyl carbons can be poor "reporter" atoms for β-^{2}H shifts and the presence of ^{2}H label was shown by direct ^{2}H NMR analysis of the

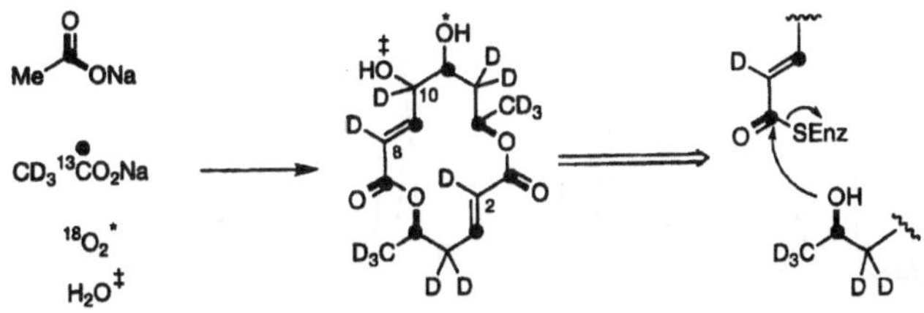

Fig. 5. Incorporation of $[1\text{-}^{13}C,^{18}O_2]$-, $[1\text{-}^{13}C, 2\text{-}^2H_3]$acetates and $^{18}O_2$ into colletodiol and an acyl substitution mechanism for macrolactonisation

enriched metabolite. The level of ^{2}H incorporation is essentially uniform at all the enriched positions except for C-10 where a very low level was observed.

From these results it can be concluded that the lactone ring formation occurs by an acyl substitution mechanism, as shown in Fig. 5 from the thioester intermediates (34) and (35) to give the macrocyclic triene (36). Examination of models suggested that the triene (36) would adopt a conformation similar to that observed for colletodiol and in this conformation epoxidation should occur from the more accessible 10Re,11Re face of the Z-alkene to give the 10S,11R-epoxide (37) which on hydrolysis would give colletodiol with the correct 10R,11R stereochemistry and the observed origins of the hydroxyl oxygens. Both the triene (36) and epoxide (37) have been prepared, specifically ^{2}H-labelled at C-10, from [10-^{2}H]colletodiol, itself prepared by reduction of colletoketol with sodium borodeuteride [30]. On feeding to *Cytospora*, both (36) and (37) result in specific incorporation of ^{2}H label at C-10 of colletodiol, proving that both the triene and the epoxide are specific intermediates. The level of incorporation of ^{2}H was lower than expected from control experiments, which is consistent with the observed low level of incorporation from ^{2}H-labelled acetate at this position and may be due to a facile redox equilibrium between colletodiol (38) and colletoketol (39). Subsequent studies have shown the presence of novel 13-membered macrodiolides bartanol (41) and bartallol as minor co-metabolites of colletodiol [31]. It may be envisaged that a concerted rearrangement of epoxide (37) would produce the aldehyde (40) which on reduction would give bartanol directly with the correct absolute stereochemistry. This appears to provide the first example of macrolides in which the ring size has been established by a ring contraction process on a preformed macrocycle. The structure of bartanol has been confirmed by an enantioselective total synthesis [32].

As indicated in Scheme 12, the labelling studies are consistent with formation of triene (36) from thioesters (34) and (35) which themselves may be built up by the assembly sequence shown. Diol (33), in which the stereochemistry at C-3 is unknown, is proposed as a common intermediate, elimination to give either an E or Z alkene leading respectively to (34) or, after further condensation, reduction and elimination, to (35). Experiments to test this sequence with the putative intermediates fed in the form of their N-acetylcysteamine (NAC) thioesters have yet to be completed, but, significantly, feeding studies in which the putative triketide (34) has been fed as the NAC thioester, as the ethyl ester or

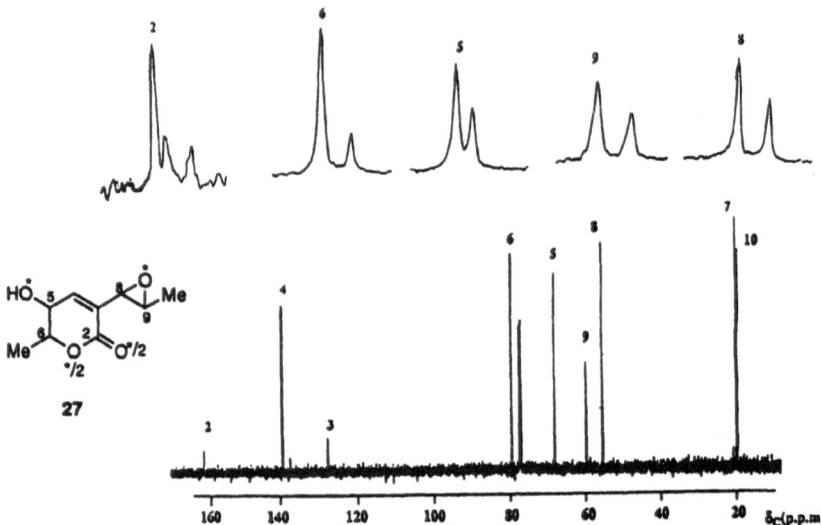

Fig. 6. ^{18}O-Isotope induced shifts in the 100.6 MHz p.n.d. ^{13}C NMR spectrum of $^{18}O_2$ enriched aspyrone (**27**)

as the free carboxylic acid have all failed to provide evidence of intact incorporation [unpublished results]. The lack of incorporation could be explained by the diol (**33**) being the immediate precursor for the macrolactonisation reaction with dehydration to triene (**36**) occurring after macrocycle formation.

The development of ^{18}O and ^2H labelling experiments led to a re-examination of the formation of aspyrone (**27**). Incorporation of $[1-^{13}C,^{18}O_2]$acetate and $^{18}O_2$ gas revealed the surprising result that none of the oxygens were derived from acetate, three being derived from the atmosphere and one from the medium [33]. As indicated in Fig. 6, the lactone carbonyl carbon C-2 showed isotope shifts due to the presence of aerobically-derived ^{18}O in either of the doubly or singly bonded oxygens. These results were subsequently verified in one of the few reported applications of ^{17}O labelling [6] and ^{17}O NMR in biosynthetic studies [34]. Although NMR active (spin 5/2), oxygen 17 is an insensitive nucleus which gives broad lines due to quadrupolar relaxation. In addition, acoustic ringing in the NMR probe and pulse breakthrough can cause serious interference with the recording of the signal and this may result in severe baseline roll which may obscure weak signals. The quality of ^{17}O NMR spectra has been greatly improved by application of the maximum entropy method [34] which does make its use in biosynthetic studies more attractive.

Incorporation of $[2-^{13}C,^2H_3]$acetate into aspyrone (**27**) gave the spectrum shown in Fig. 7. As may be seen, simultaneous deuterium and proton noise decoupling leads to a great simplification of the otherwise uninterpretable results for C-7 and C-10. The retention of two acetate-derived hydrogens at C-7 makes it unlikely that this carbon could have been part of an olefinic bond at any

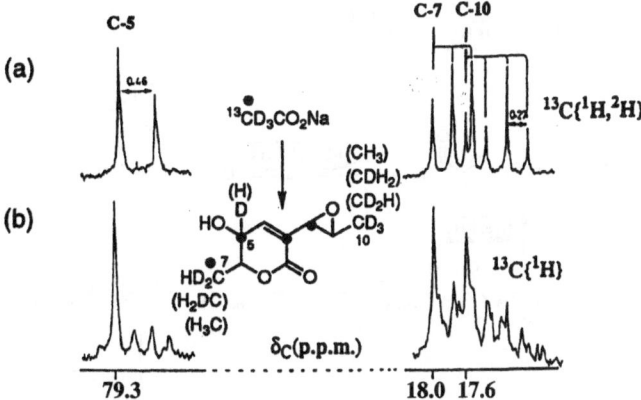

Fig. 7 a, b. α-²H Isotope induced shifts observed in the 100.6 MHz ¹³C NMR spectra of aspyrone (27) enriched from feeding [2-¹³C,²H₃]acetate to *Aspergillus melleus*: **a** with simultaneous ¹H and ²H noise decoupling; **b** with ¹H noise decoupling only

point in the biosynthetic pathway. To accommodate these results, a pathway involving epoxide-mediated rearrangement and ring-closure reactions was proposed. This was further modified by a spectacular series of incorporation experiments with putative chain assembly intermediates which were synthesised in the form of their NAC thioesters with either ²H or double ¹³C labels [35–40]. These experiments resulted in a complete delineation of the assembly pathway catalysed by the PKS and remain the most complete and successful study of its type. Incorporation of β-hydroxythioester intermediates, e.g. the NAC thioester corresponding to (42) [38, 39], established that the ketoreduction process occurs with the same absolute stereochemistry as in fatty acid biosynthesis. This led to the proposal that the PKS involved in aspyrone biosynthesis could be derived from a fatty acid synthase which presumably had lost the capacity to carry out enoyl reductions. The dehydrations were shown to occur by a *syn* elimination [40]. The final hexaketide product (43) of the polyketide assembly process would be converted to aspyrone by the sequence shown in Scheme 13 which includes an analogous epoxide rearrangement to that proposed for bartanol biosynthesis above.

While some success has been reported in analogous studies with polyketide assembly intermediates in *Streptomyces* metabolites, e.g. erythromycin [41] and tylosin [42], similar experiments on fungal polyketides have been more limited. The di- and tetraketide intermediates (44) and (45), variously doubly labelled with ¹³C and ¹⁸O as indicated in Scheme 14, have been incorporated into dehydrocurvularin (46) by cultures of *Alternaria cineriae* [43]. However, in contrast to the ease of incorporation of assembly intermediates into aspyrone by *A. melleus*, the experiments in *A. cineriae* required considerable experimentation to optimise the feeding conditions and the use of the β-oxidation inhibitors. The initial experiments [43] depended on the use of UV mutants of *A. cineriae* which had lost the ability to utilise fatty acids and therefore to degrade the fatty

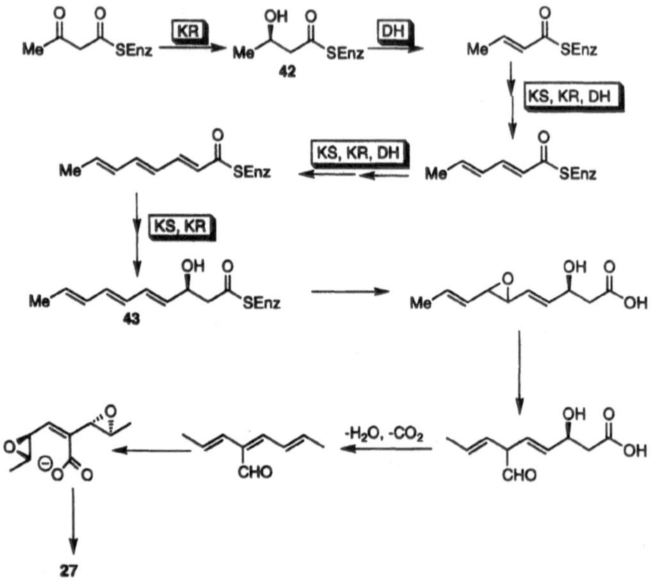

Scheme 13

acid-like labelled precursors. In a subsequent paper [44] it was reported that a range of β-oxidation inhibitors, and in particular 3-(tetradecylthio)propanoic acid, were effective in allowing intact incorporation to be observed. Incorporations of up to 70% were obtained, including that of ethyl (7S)-[6,7-$^{13}C_2$,*hydroxy-*^{18}O]-7-hydroxyoct-2-enoate. The ^{13}C NMR spectrum of the enriched dehydro-curvularin showed an ^{18}O isotopically shifted doublet for C-4 and a simple doublet due to ^{13}C-^{13}C coupling with no isotopic shift for C-5.

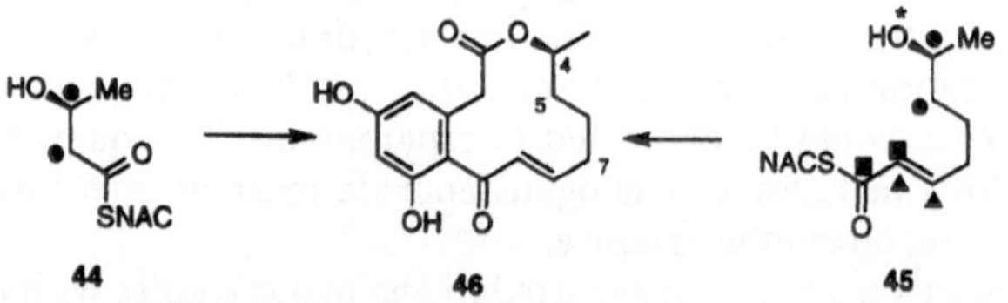

Scheme 14

Our own studies on the biosynthesis of monocerin (50), Scheme 16, in *Dreschlera ravenelii* illustrate the problems which commonly arise in this type of study. Incorporations of ^{13}C, ^2H and ^{18}O-labelled acetates and $^{18}O_2$ gas and analysis by ^{13}C and ^2H NMR showed inter alia that the oxygen atoms attached to C-9 and C-11 were acetate derived, so that successive ketoreductions occur with opposite stereochemistry; that both hydrogens at C-10 are acetate derived, consistent with ketoreduction occurring during chain assembly; and that the 'extra' oxygen at C-4 is derived aerobically [45]. These results are consistent with

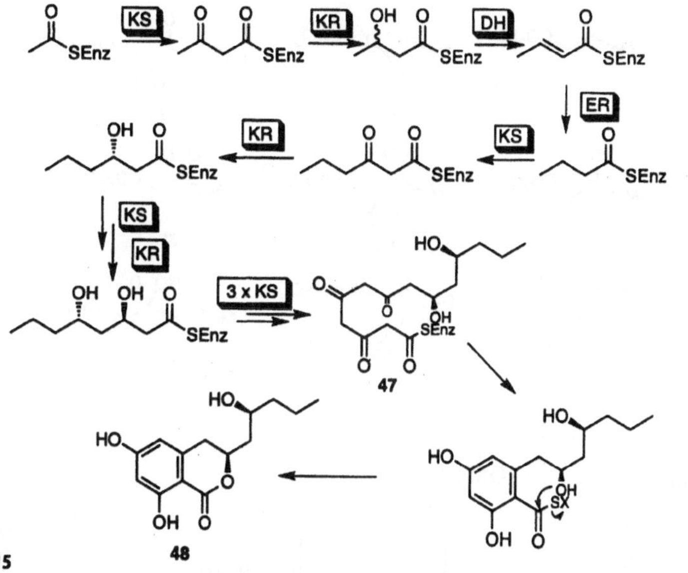

Scheme 15

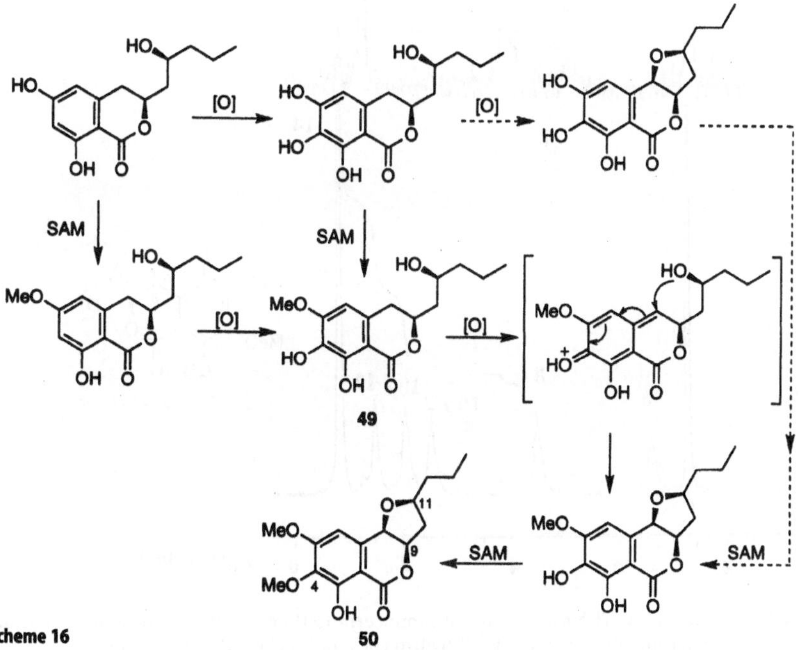

Scheme 16

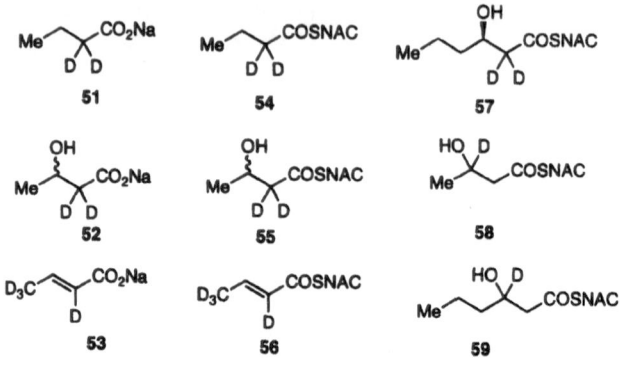

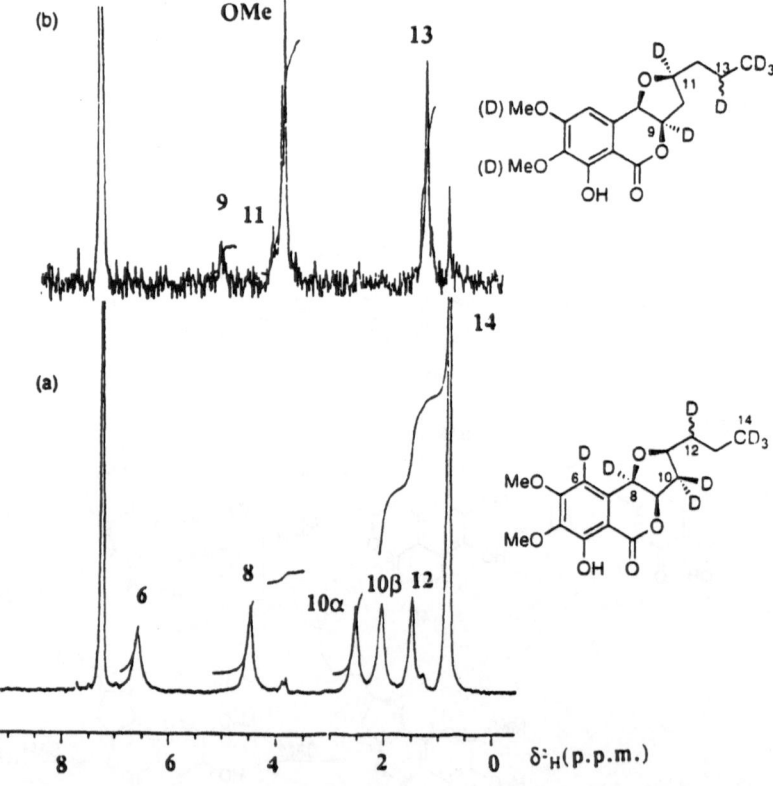

Fig. 8 a, b. 55.28 MHz ²H NMR spectra of monocerin (50) enriched from feeding: **a** sodium [2, 4, 4, 4-²H₄]crotonate (53); **b** [3-²H]-3-hydroxyhexanoic acid NAC thioester (59)

the polyketide assembly sequence shown in Scheme 15 to give the heptaketide (47) as the product of the assembly phase. This would cyclise and aromatise and finally lactonisation of the C-9 hydroxyl onto the thioester would give the dihydroisocoumarin (48) as the first PKS-free intermediate. This would be further converted into the final metabolite as outlined in Scheme 16, in which the exact sequence of the necessary modification steps is unclear. Support for this sequence does come from the isolation of the known fusarentin ether (49) as a co-metabolite of monocerin. In order to establish the exact sequence, an enantioselective synthesis of the fusarentins which will allow preparation of the necessary di- and trioxygenated dihydroisocoumarin intermediates doubly labelled with ^{13}C has been developed [46].

A number of putative assembly intermediates have been prepared in ^{2}H labelled forms for feeding studies to confirm the assembly sequence shown in Scheme 15. However, none of these were incorporated intact by cultures of *D. ravenelii*, although all were efficiently metabolised by the cultures [54]. The results fell into three main categories. The sodium salts (51)–(53) all showed high levels of ^{2}H incorporation by ^{2}H NMR, which corresponded exactly to that observed for incorporation of [^{2}H$_3$]acetate (Fig. 8a). The corresponding thioesters (54)–(57) gave a similar labelling pattern, but the incorporation levels were considerably lower. Interestingly, the thioesters (58) and (59) gave efficient incorporation of ^{2}H label at those positions, including the methoxyls, where one would predict incorporation of hydrogen by reductive modification (Fig. 8b). It appears therefore that these precursors are broken down by β-oxidation with concomitant enrichment of the NADPH pool, and that the resultant NADPD is immediately used in the biosynthetic pathway to monocerin.

4
Stereochemistry of Polyketide Assembly in Fungi: ^{13}C, ^{2}H and ^{3}H Studies

Many of the above studies have given invaluable information on the stereochemical outcome of the ketoreductase and dehydratase catalysed reactions occurring during polyketide assembly in fungi. A number of studies of incorporation of [^{2}H$_3$]acetate have provided indirect information on the stereochemistry of the final enoyl reductase reaction. Thus ^{2}H label is found at the pro-S positions at C-11 of cladosporin (60) in *Cladosporium cladosporoides* [47], C-7 of dehydrocurvularin (46) and C-8 of antibiotic A26771B (61) in *Penicillium turbatum* [48]. In all cases these results indicate that the enoyl reductase of the PKS has the opposite stereochemistry to that of the corresponding FAS. In contrast to the above, the ^{2}H labels at C-2' and C-4' of averufin (62) in *Aspergillus parasiticus* show the same stereochemistry as those in the corresponding fatty acids [49]. This result is consistent with the proposed [50] intermediacy of a hexanoate starter unit from fatty acid metabolism in averufin biosynthesis. Recent genetic evidence has indicated that there is a specialised FAS associated with the PKS involved in averufin biosynthesis and in some elegant experiments it has been shown that exogenous hexanoyl NAC could be effective in substituting for the product of the FAS when the gene is disrupted [51].

The stereochemistry of the ketosynthase-catalysed condensation of malonate with the enzyme-bound thioester at the start of each elongation cycle has

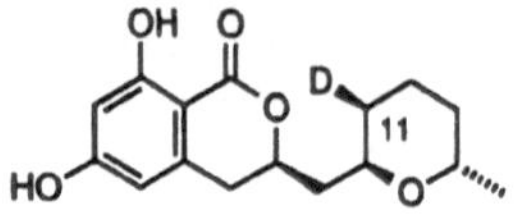

60 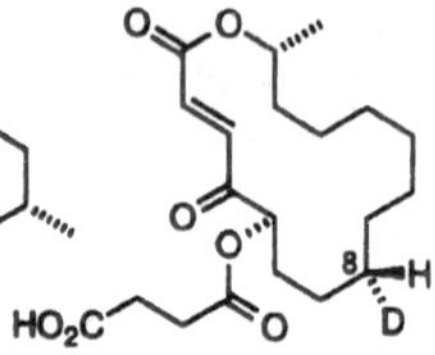 **61** 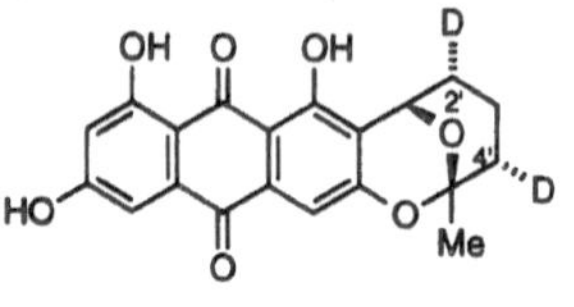 **62**

always been presumed to occur with inversion of configuration at malonate as has been established for the corresponding condensation in fatty acid biosynthesis. In collaboration with Heinz Floss [unpublished results], we have obtained information for this from the incorporation of (2R)- and (2S)-[^{3}H,^{2}H,^{1}H]acetates into colletoketol (**39**) and monocerin (**50**) and subsequent ^{3}H NMR analysis of the enriched metabolites.

Although best known as a radioactive tracer the ^{3}H nucleus has ideal characteristics for use as an NMR label. As the natural abundance is practically zero and the magnetogyric ratio is the highest for any nuclide (28.53 compared to 26.75 for ^{1}H), ^{3}H NMR spectroscopy is uniquely sensitive by the standard of other NMR methods of tracing isotopes. The chemical shift values and coupling constants are very close to those of the corresponding ^{1}H NMR spectrum, so assignments can be made on this basis. Accurate integration is also possible. Tritium NMR has been used in a number of whole cell and enzyme studies [6] including study of the incorporation of ^{3}H-labelled valines into cephalosporin C in *Cephalosporium acremonium* [52], and incubation of [^{3}H]porphobilinogen with porphobilinogen deaminase from *Rhodopseudomonas* sp. [53].

As indicated in Scheme 17 for the R-acetate, assuming that carboxylation to malonate proceeds with retention of configuration, there are four possible tritiated products from the condensation step. Thus on subsequent incorpora-

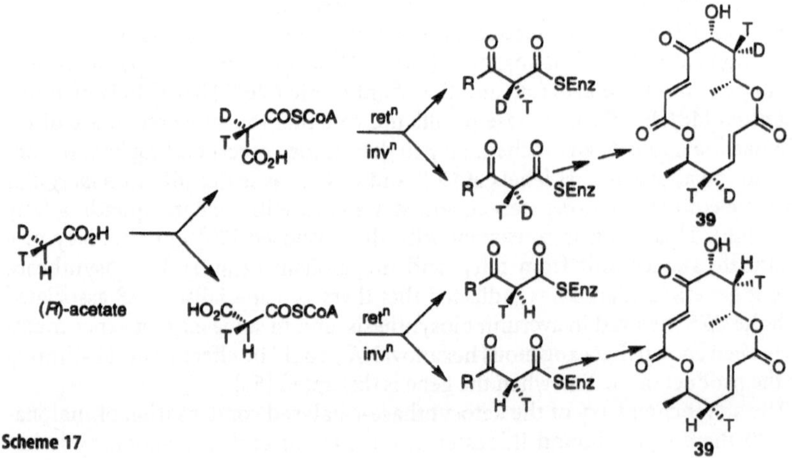

Scheme 17

39

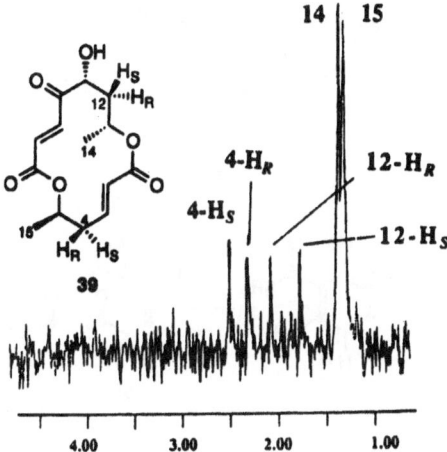

Fig. 9. 533 MHz ^{3}H NMR spectrum of colletoketol (**39**) enriched from feeding sodium (R)-[^{3}H,^{2}H,^{1}H]acetate (50 mCi, 36 mCi mmol^{-1}) to *Cytospora* sp. ATCC 20 502

tion into the 4- and 12-methylene groups of colletoketol (**39**) or the 10-methylene group of monocerin (**50**), all positions where the stereospecific assignments of the diastereotopic hydrogens have been made, the stereochemical outcome can be deduced once it is known which of the prochiral hydrogens is incorporated along with ^{2}H and which is incorporated along with ^{1}H. This should be apparent from analysis of the p.n.d. ^{3}H NMR spectrum in which one ^{3}H signal should be broadened by coupling to ^{2}H. The results for incorporation of (R)-acetate into colletoketol in *Cytospora* sp ATCC 20502 are shown in Fig. 9. ^{3}H Label is incorporated most effectively into the methyl positions, but there is significant incorporation into all four methylene hydrogens. Interestingly, there is no observable incorporation into the 2- or 8-alkene positions, in contrast to the results when [^{2}H$_3$]acetate is fed. This is difficult to rationalise, but it may be due to a "pool" effect, with the small amount of labelled acetate used in the experiment being entirely consumed by the PKS for the first two condensations, with unlabelled endogenous acetate being used for subsequent condensations.

Close examination of the spectrum reveals a sharp singlet at $\delta 2.54$, indicating ^{3}H label at the 4-pro-S position with protium adjacent and a broader signal at $\delta 2.35$, indicating ^{3}H at the 4-pro-R position with deuterium adjacent. This is consistent with inversion of configuration. The results for the 12-methylene group is less clear as both the pro-R and pro-S hydrogen at $\delta 2.14$ and 1.85 appear as sharp singlets. The results for the 10-methylene group in monocerin are, however, also consistent with inversion of configuration.

No discussion of studies on the stereochemical outcome of polyketide biosynthesis would be adequate without mention of the seminal studies of Jordan and Spencer [56–58] using chiral malonates to determine the stereospecificity of the elimination and enolisation reactions occurring during 6-MSA (**66**, R=H) and orsellinic acid (**66**, R=OH) biosynthesis. Early experiments by Abell and

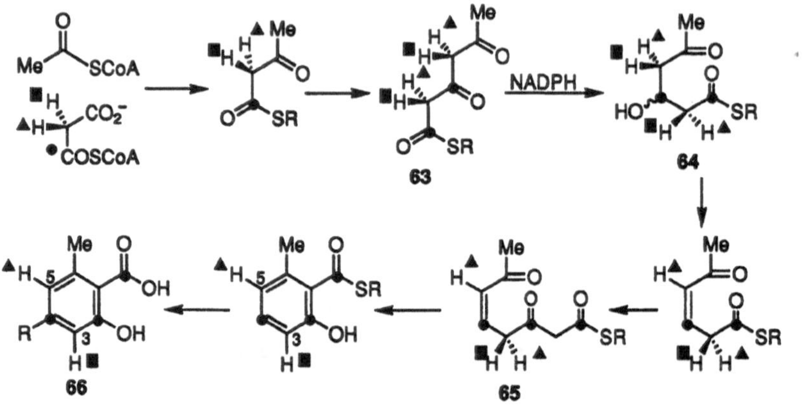

Scheme 18

Staunton in which they compared the **relative** levels of ^{2}H incorporation at the 3- and 5-positions from [2-^{2}H$_3$]- and [2-^{2}H$_1$]acetates had already indicated that the loss of hydrogen was stereospecific, but were not able to define the stereochemistry [55].

Scheme 18 illustrates the proposed stages in 6-MSA biosynthesis in which the first and second condensation steps proceed with inversion to give the triketide (63). Ketoreduction gives the alcohol (64) and then elimination followed by a final malonyl condensation generates the tetraketide (65) which cyclises via an intramolecular condensation and enolises to give the aromatic nucleus of (66). In the first set of experiments (R)- and (S)-[1-^{13}C,^{2}H]malonates were incubated separately with 6-MSA synthase purified from *Penicillium patulum* [56]. Isotope incorporations were determined by mass spectrometry. All the possible isotope patterns for retention or loss of the pro-R or pro-S hydrogens from C-3 and C-5 were permutated. Comparison with the actual spectra obtained demonstrated that opposite prochiral hydrogens were eliminated. The absolute stereochemistry was established in an analogous experiment [57] where the chiral malonates were incubated with acetoacetyl CoA rather than acetyl CoA. Subsequent mass spectral analysis showed that it is the H$_R$ proton that is retained at C-3 of 6-MSA and so it can be deduced that the hydrogen at C-5 must be derived from the opposite prochiral hydrogen, H$_S$. The overall result is summarised in Scheme 18. In a recent collaborative study we have synthesised the triketide alcohol (64) as its NAC thioester and shown that it is indeed a precursor as, on incubation with 6-MSA synthase and malonyl CoA, 6-MSA production is observed [unpublished results]. Current work is aimed at synthesis of both enantiomers of (64) to study the overall stereochemistry of the ketoreduction and elimination reactions.

A similar stereochemical outcome to that observed for 6-MSA was found using orsellinate synthase isolated from *Penicillium cyclopium* [58]. The hydrogens retained at C-3 and C-5 of orsellinic acid (66, R = OH) are from opposite prochiral sites in malonate. This contrasts with a previous study [59] on orsellinic acid carried out in the same way which had concluded that the hydrogen loss was non-stereospecific. The difference may be ascribed to an unfavourable rate of deuterium isotope exchange relative to the rate of orsellinic acid synthesis, leading to racemisation in the latter experiment.

5
Incorporation of Fluorinated and Other Substrate Analogues

There are many examples of polyketide metabolites where the common acetate starter unit is replaced by other groups, and a general discussion of these is outside the scope of this review. Numerous studies have been carried out on the use of analogues of the normal biosynthetic precursors to produce novel structures. This type of "precursor-directed biosynthesis" has been the subject of a recent comprehensive review [60].

A number of recent studies have investigated the use of fluorine as a biosynthetic label. Fluorine is of interest because of its relatively low cost in comparison with ^{13}C-labelled compounds and the power of ^{19}F NMR spectroscopy which is of comparable sensitivity to ^{1}H NMR. Moreover, efficient incorporation of a fluorinated precursor and isolation of the resultant novel metabolite is a potential method for generation of analogues for biological testing. Fluorine is the most electronegative element, resulting in a strongly polarised C-F bond, but it has similar steric requirements to hydrogen. Thus, the electronic consequences of replacing hydrogen are significant, affecting substrate-enzyme binding affinity and, ultimately, the biological activity of the fluorinated analogue [61]. The squalestatins, e.g. (68), are novel inhibitors of squalene synthase and hence of cholesterol biosynthesis. They were isolated from a *Phoma* species and extensive studies with ^{13}C, ^{2}H and [18]O labelled precursors established that the dioxabicyclooctane core was formed via condensation of a hexaketide precursor primed with a benzoyl CoA starter with a Krebs' Cycle intermediate, presumably oxaloacetate [62]. In an attempt to produce novel analogues for biological screening, a variety of halogenated, alkylated and other substituted benzoic acids were fed to squalestatin-producing cultures. Incorporation was successful with a few of the fluorinated analogues, e.g. (69) and (70), and incorporation levels of up to 33% were observed [63] as indicated in Scheme 19.

In another interesting study on the biosynthesis of the antibiotic tetronasin (71), various di-, tri- and tetraketide substrate analogues were synthesised [64, 65] as their NAC thioesters and fed to *Streptomyces longisporoflavus* [66]. The analogues were different by virtue of the fact that one of the natural pendant or terminal methyl groups was replaced by an ethyl, *iso*-propyl or benzyl moiety. In addition to this, the α-hydrogen was replaced by fluorine in several of the substrates. The synthesis of these precursors involved an interesting application of electrophilic fluorinating agents to Evans' oxazolidinone chemistry and serves to highlight again the necessity for carrying out demanding synthetic work in many biosynthetic studies.

The introduction of fluorine had several purposes. These were to provide a biosynthetic label for detection of incorporation by ^{19}F NMR; to produce structural analogues of the natural metabolite; and specifically in this case to inhibit

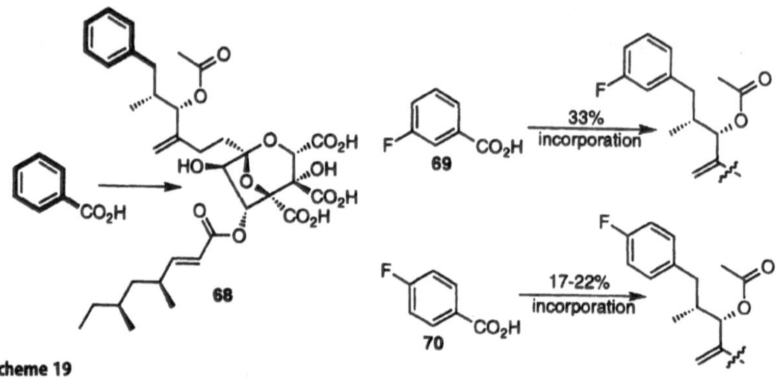

Scheme 19

the β-oxidation process which commonly results in degradation of these assembly intermediates and prevents their intact incorporation. Examples of two tetraketide analogues (72) and (73) from the extensive range of modified precursors tested are shown in Scheme 20 along with the expected novel tetronasin analogues (74) and (75). The generation of novel fluorinated analogues was not efficient enough to be detected by ^{19}F NMR. However, in an interesting application of high resolution electrospray mass spectrometry, $[M+K]^+$ peaks could be seen in electrospray mass spectra of partially purified extracts from the incorporation experiments and the levels of incorporation estimated by comparison with the relevant $[M+K]^+$ peak of the natural metabolite (Fig. 10). In this way, levels of incorporation as small as 0.1 % could be readily detected. It was interesting, but perhaps not surprising, that incorporation was observed only with precursor analogues containing fluorine and/or where the methyl was

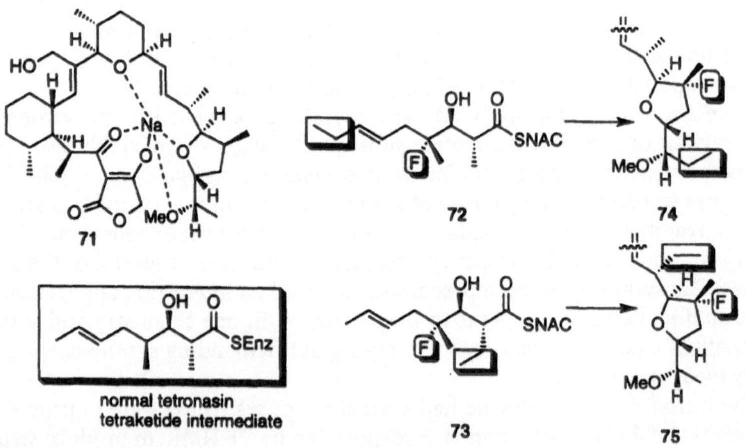

Scheme 20

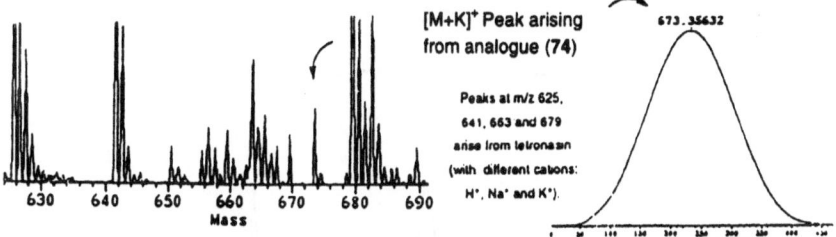

[M+K]⁺ Peak arising
from analogue (74)

Peaks at m/z 625,
641, 663 and 679
arise from tetronasin
(with different cations:
H⁺, Na⁺ and K⁺).

Fig. 10. High resolution electrospray mass spectrum of tetronasin (**71**) containing analogue (**74**) resulting from feeding of precursor (**72**). (Figure kindly supplied by J. Staunton)

replaced by ethyl. The larger structural analogues, i.e. those with benzyl and *iso*-propyl groups, were not incorporated. This indicates that the PKS has some, but limited, relaxation of substrate specificity for assembly intermediates.

The success, albeit limited, of incorporation studies of polyketide assembly intermediates has resulted from feeding these in the form of their NAC thioesters which structurally mimic the thiol end of the phosphopantetheine moiety found in coenzyme A and the acyl carrier protein component of the PKS. This will be discussed further below, but it has also been shown that there are advantages to feeding starter units in the form of their NAC thioesters.

There is considerable indirect and direct evidence that the PKS responsible for the assembly of norsolorinic acid (NSA, **79**), the first isolable intermediate in the pathway to the aflatoxins, is primed by a hexanoate starter and, indeed, as discussed above, feeding [1-¹³C]hexanoic acid to averufin-producing cultures of *Aspergillus parasiticus* was reported to give some intact incorporation. To study this further we developed a method for production of NSA (**79**) in shake cultures and fed [2-²H₃]hexanoate (Scheme 21) to these cultures in the form of the free acid (**76**), the ethyl ester (**77**) and finally the NAC thioester (**78**) [67]. ²H NMR

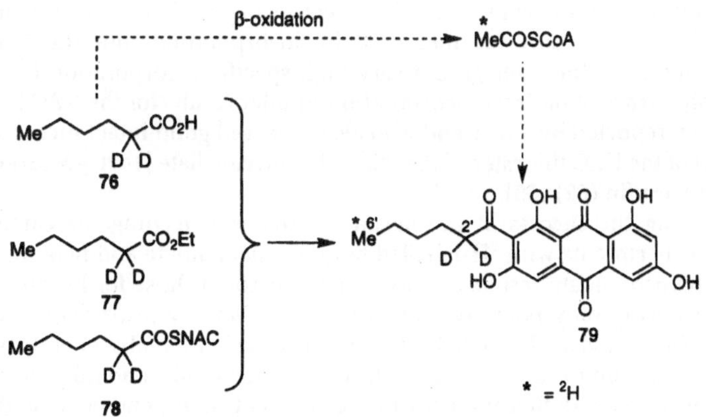

Scheme 21

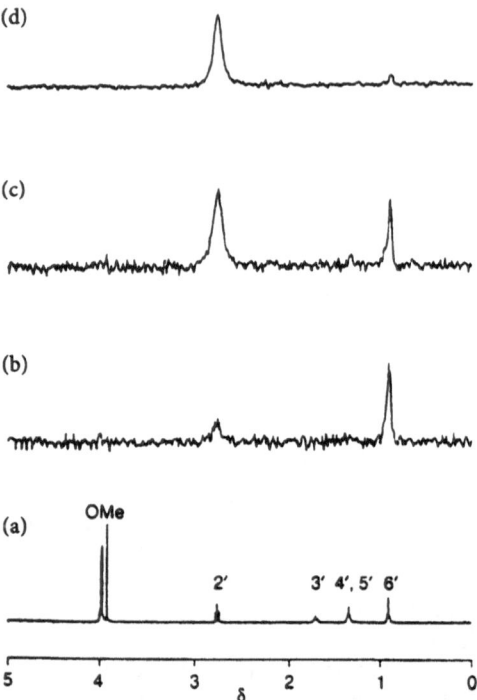

Fig. 11 a–d. 0–5 Ppm regions of: **a** the 400 MHz ^{1}H NMR spectrum of norsolorinic acid (**79**) as its tetramethyl ether; **b** the 62 MHz ^{2}H NMR spectrum enriched from feeding [2-^{2}H$_2$]hexanoate in the form of the free acid (**76**); **c** the 62 MHz ^{2}H NMR spectrum enriched from feeding [2-^{2}H$_2$]hexanoate in the form of the ethyl ester (**77**); **d** the 62 MHz ^{2}H NMR spectrum enriched from feeding [2-^{2}H$_2$]hexanoate in the form of the NAC thioester (**78**)

analysis of the NSA isolated in each case is shown in Fig. 11. It can be seen that whereas the free acid gave a low direct incorporation of label into the 2′-methylene against a high background of indirect incorporation via degradation to ^{2}H-labelled acetyl CoA (as measured by incorporation into the 6′-methyl group), the NAC thioester gave a very high specific incorporation (ca. 45%) with only a trace of indirect incorporation. Similar results for the NAC thioester have been reported by Townsend who also observed good intact incorporation (4–5%) of the NAC thioester of the "diketide" intermediate [1-^{13}C]-3-oxooctano-ate into averufin (**62**) [50].

Following the success of the hexanoate thioester feeding, we carried out similar experiments with ^{2}H-labelled butyrate, pentanoate and heptanoate. Of these, the pentanoate results almost exactly mirrored those for hexanoate, the others showing only poor incorporation exclusively by prior degradation to acetate. The pentyl analogue (**80**) of norsolorinic acid was subsequently isolated by preparative hplc separation from the natural metabolite and fully characterised. Thus it appears that the NSA PKS can accept both pentanoate and hexanoate starters with comparable facility.

To probe the substrate specificity further, the incorporation of the NAC thioester of 6-fluorohexanoate was examined. The 470 MHz ^{19}F NMR spectrum of the metabolite (derivatised as its tetramethyl ether) showed a very strong triplet of triplets at δ –218.2 ppm, similar to that observed in the thioester. The mass spectrum showed a large amount of a new metabolite of MW 444 (Fig. 12b). In the 500 MHz ^{1}H NMR spectrum (Fig. 12a), the 6'-methylene signal of the 6'-fluoroanalogue (81) appeared as a doublet of triplets, whereas the

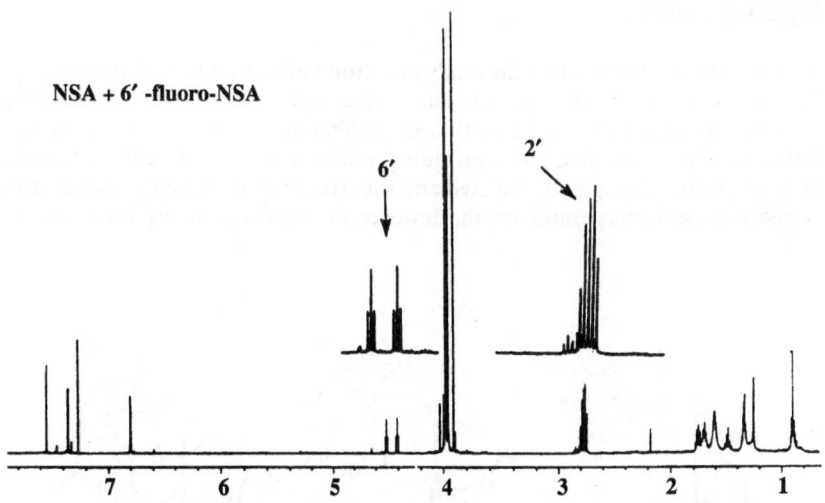

6'-methyl signal for the co-produced natural metabolite appeared as a triplet at 0.90 ppm. Integration of the signals indicated that the fluoroanalogue comprised ca. 40% of the mixture which again was confirmed by hplc separation. In contrast, when the thioester of 2-fluorohexanoic acid was fed, there was no evidence whatsoever of fluorine incorporation and the yields of NSA were severely reduced, suggesting that the thioester may be acting as an inhibitor of the PKS.

Fig. 12 a. 500 MHz ^{1}H NMR spectra of the tetramethyl ethers of norsolorinic acid (79) and the 6-fluoro-analogue (81) resulting from feeding 6-fluorohexanoic acid as its NAC thioester

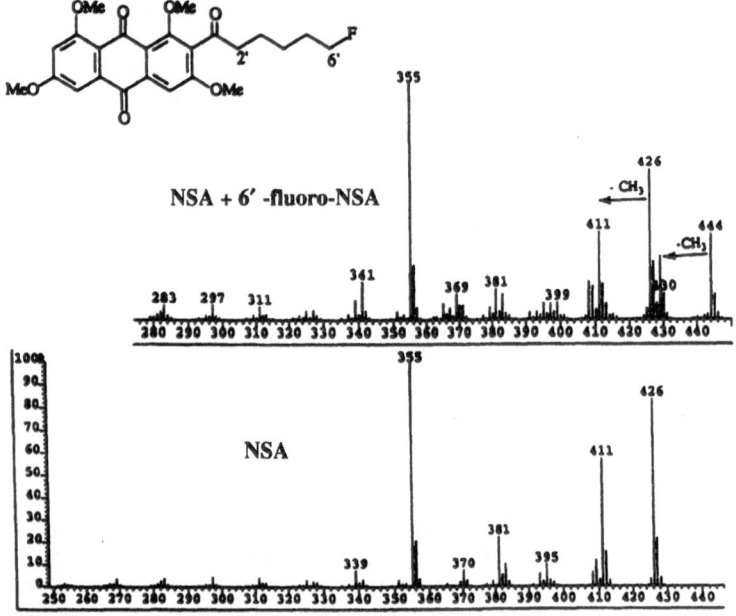

Fig. 12 b. EI mass spectra of the tetramethyl ethers of norsolorinic acid (**79**) and the 6-fluoro-analogue (**81**) resulting from feeding 6-fluorohexanoic acid as its NAC thioester

6
Meroterpenoids

A major interest has been in the study of a group of compounds of mixed poly-ketide-terpenoid origins – the so-called meroterpenoids. Our interest in these compounds began with the isolation of andibenin B (**82**) from *A. variecolor*. Other clearly related structures including andilesin A (**83**) [68] and anditomin (**84**) [69] were subsequently isolated and the structures of these C_{25} metabolites suggested a sesterterpenoid origin. However, the results of incorporation expe-

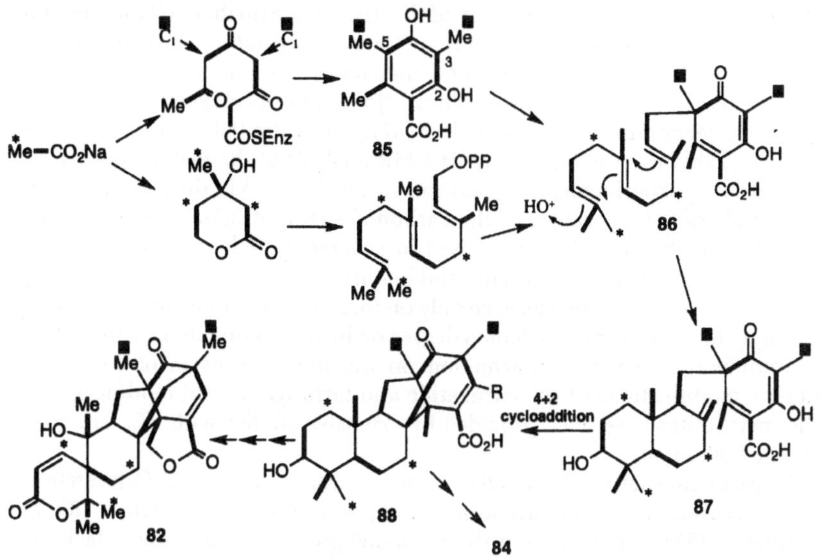

Scheme 22

riments [70] with ^{13}C-labelled acetates and methionine gave a labelling pattern indicative of a mixed polyketide-terpenoid biosynthesis (Scheme 22) in which a di-C-methylated tetraketide-derived phenolic precursor (85) is alkylated by farnesyl pyrophosphate to give the cyclohexadienone (86), which after epoxide-mediated cyclisation to the triene (87) undergoes an intramolecular Diels-Alder reaction to generate the bicyclo[2.2.2]octane system (88) observed in andibenin B and andilesin A. Subsequent oxidative modifications and elaboration of the terpenoid portion would generate the ε-lactone and spiro-δ-lactone systems of (83) and (82) respectively, whereas a carbocation initiated ring expansion of the carbocyclic ring of the tetraketide-derived moiety in (88) would generate the rearranged ring system observed in anditomin (84). While this work was in progress, the isolation was reported of two further metabolites for which sesterterpenoid origins were proposed. These were the mycotoxins austin (89) and terretonin (90) from *Aspergillus ustus* [71] and *Aspergillus terreus* [72] respectively. Similar incorporation experiments with ^{13}C-labelled acetates and

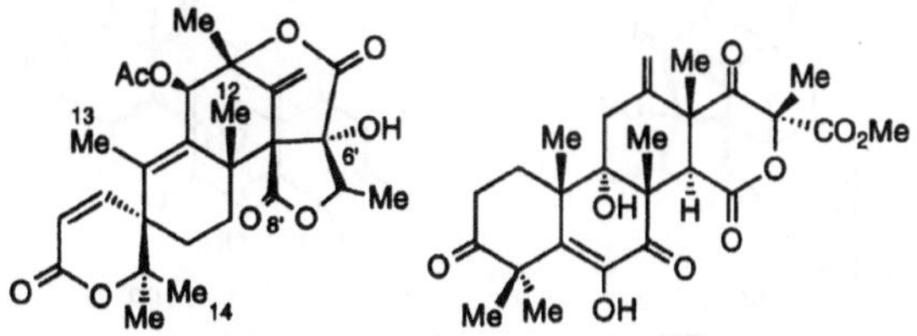

(89)   **(90)**

methionine gave results that suggested that these were further metabolites of the meroterpenoid pathway via the common intermediate (86). Conclusive evidence for the meroterpenoid origins of these metabolites was obtained by the synthesis of ²H-labelled 3,5-dimethylorsellinic acid (85) (as its ethyl ester) [73] and its specific incorporation into andibenin B (82) [74], austin (89) [75] and terretonin (90) [76]. This was established by ²H NMR analysis of the metabolites isolated from feeding experiments with (85) specifically labelled in the 3-methyl group. The mode of incorporation of the carbon skeleton of 3,5-dimethylorsellinate into these metabolites is summarised in Scheme 23. Whereas the carbocyclic ring is incorporated intact into the andibenins (e.g. 82) and andilesins (e.g. 83), and undergoes one-bond cleavage only on incorporation into anditomin (84), it is fragmented to an unprecedented degree on incorporation into austin (89) and terretonin (90). Further information on the mechanisms involved in these drastic modifications of the tetraketide and farnesyl derived modifications of the metabolites have been provided by extensive studies with ¹³C, ²H and ¹⁸O labelled precursors.

The labelling pattern obtained from incorporations of [1-¹³C,¹⁸O₂]acetate and ¹⁸O₂ were consistent with formation of the γ-lactone in the andibenins (82) and andilesins (83) by oxidation of the 1'-methyl group and subsequent lactonisation on to the C-8' carboxyl [77, 78]. The results of incorporation of ¹⁸O₂ into austin (89) indicated a modification mechanism in which the orsellinate moiety undergoes a ring contraction via an α-ketol rearrangement followed by Baeyer-Villiger-type oxygen insertion to form the δ-lactone as shown in Scheme 24. Low incorporation levels precluded the observation of isotope shifts from [1-¹³C,¹⁸O₂]acetate into austin, but this was overcome by synthesising [79] 3,5-dimethylorsellinate doubly labelled with ¹³C and ¹⁸O in both the carboxyl

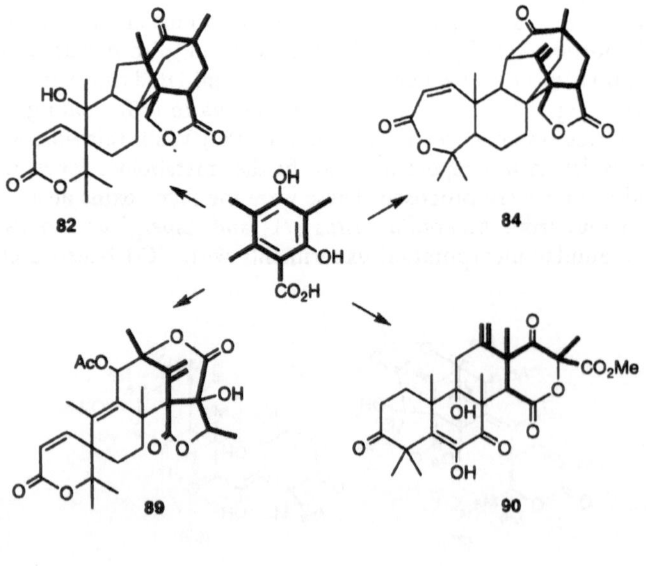

Scheme 23

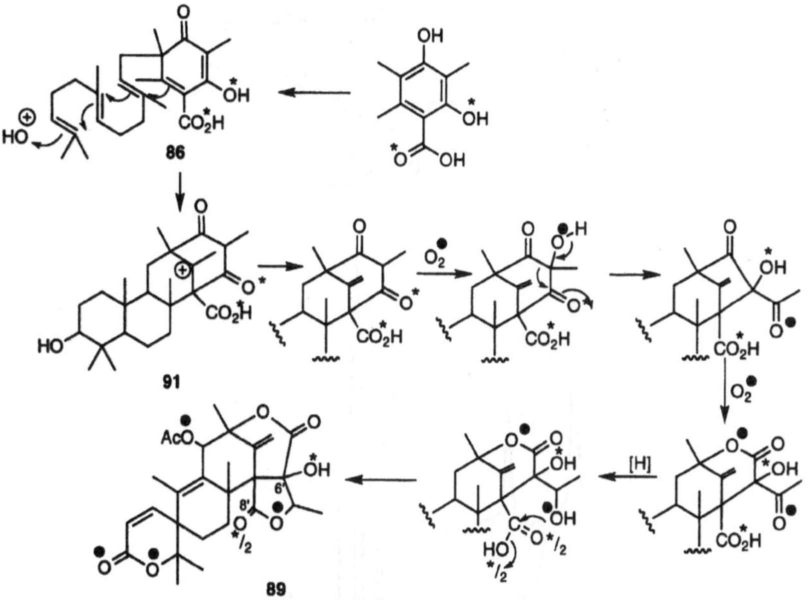

Scheme 24

carbonyl and at the C-2 position. This was incorporated with high efficiency into austin (Fig. 13), confirming the $^{18}O_2$ results [75]. The level of incorporation of ^{18}O label into the C-8' carbonyl is approximately half that into the C-6' hydroxyl, consistent with γ-lactone formation via the free carboxyl, as shown in Scheme 24.

Similar labelling experiments [76] were consistent with the formation of terretonin (90) via the same tetracylic carbocation (91) as proposed in the biosynthesis of austin (89). Ring contraction as indicated in Scheme 25 would generate the cyclopentadione (92). Subsequent aerobic hydroxylation, lactonisation and carbon-carbon bond cleavage via a retro-Claisen reaction gives the intermediate carboxylic acid (93). Methylation and oxidative modification of the terpenoid moiety would give the final structure (90). The subsequent isolation of the citreohybridones (e.g. 104) and andrastatins (see below) gives added weight to this hypothesis.

^{18}O and ^2H labelling has also provided information on the mechanism of formation of the spiro-lactone systems observed in the andibenins (e.g. 82) and austin (89) [80]. The mass spectrum of andibenin B (82) isolated from the $^{18}O_2$ labelling experiment showed a major parent ion peak at m/z=434, indicating the simultaneous incorporation of four ^{18}O atoms. The ^{13}C NMR spectrum showed that in addition to the γ-lactone ether oxygen, the oxygens attached to C-3, C-4 and C-10 are all derived from the atmosphere. These results show that the C-3 lactone function must be formed by a biological Baeyer-Villiger type oxidation of a corresponding 3-keto precursor (94) (Scheme 26).

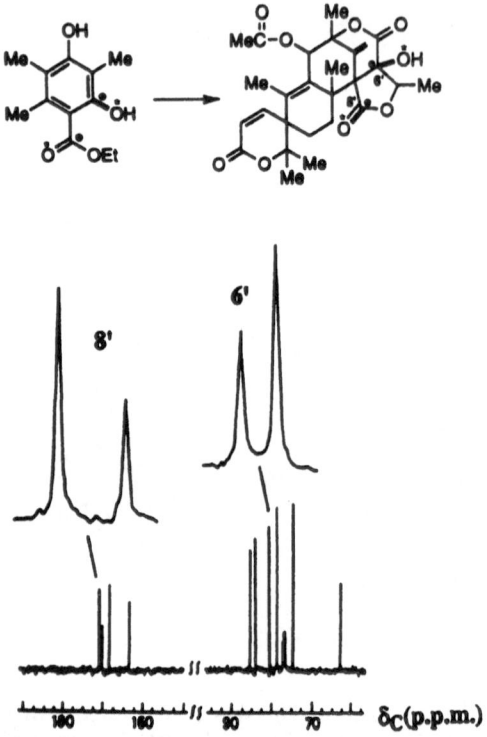

Fig. 13. ^{18}O-Isotopically shifted resonances in the 100.6 MHz p.n.d. ^{13}C NMR spectrum of austin (**89**) enriched by feeding [^{13}C,[18]O]-3,5-dimethylorsellinate to *Aspergillus ustus*

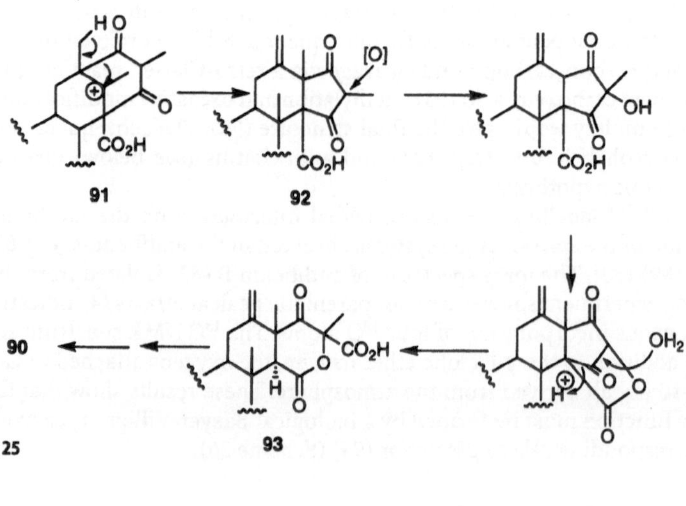

Scheme 25

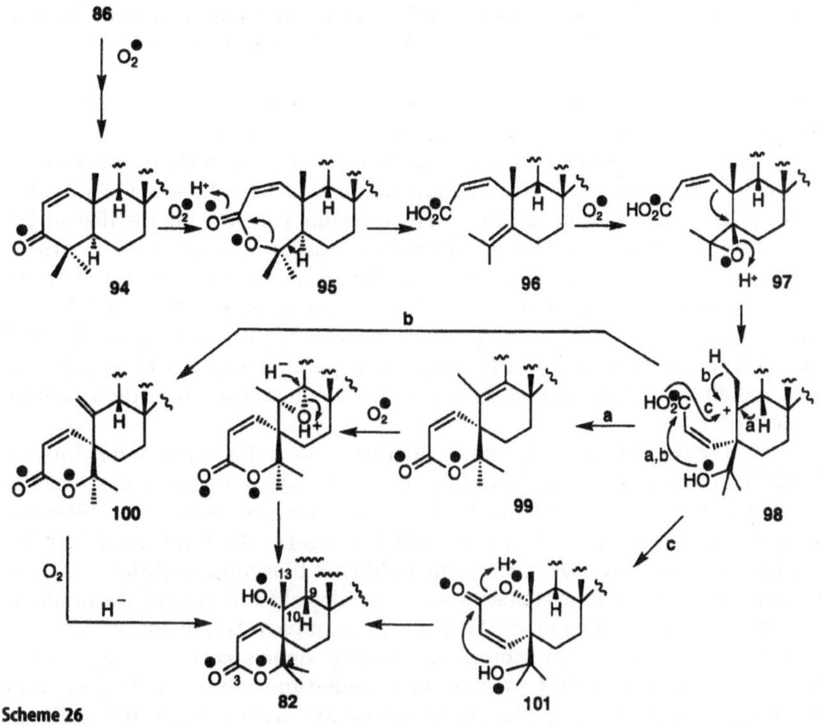

Scheme 26

The presence of atmospheric oxygen at both C-4 and C-10 suggests possible mechanisms for the biosynthesis of the spiro-lactone system. As the andilesins (e.g. 83) possess a 7-membered lactone ring, it seems reasonable to propose that the required skeletal rearrangement to form the spiro-lactone occurs subsequent to the introduction of the C-4 oxygen atom. A relatively stable carbocation could be generated from the andilesin lactone ring (95) by alkyl-oxygen cleavage. Loss of the C-5 proton would then give the tetrasubstituted alkene (96). Subsequent oxidation of (96) to epoxide (97) followed by rearrangement would give the carbocation intermediate (98). Three paths for the conversion of (98) into the spirolactone (82) can be envisaged. Elimination to the endocyclic or exocyclic alkenes (99) and (100) and subsequent epoxidation and hydride reduction (paths a and b) would account for the origin of the 10-hydroxyl in atmospheric oxygen. Any subsequent mechanism for the formation of the δ-lactone will result in two atoms of atmospheric oxygen in the lactone. The structure of austin (89) which contains both the spiro-lactone and a 9,10-double bond provides some support for this general route. An alternative route (path c) involves trapping of the C-10 carbocation (98) by the carboxyl group to form lactone (101) which then undergoes transesterification, giving the A and B rings of andibenin B directly. In this case it is not necessary to invoke a further oxidation step. Prior hydrolysis of lactone (101) to form the spiro-lactone would imply a loss of half the oxygen label at C-3 which is not consistent with the relatively similar levels of incorporation of atmospheric oxygen observed at C-3, C-4 and C-10.

These possible mechanisms have been distinguished by a series of experiments in which [5-²H₂]-, [6-²H₃] and [5-¹³C, 4-²H₂]mevalonates were incorporated into andibenin B (82) [80]. Most importantly, these experiments showed the retention of three deuterium atoms at C-13 and one at C-9 from the appropriately labelled mevalonates. These results rule out intermediates (99) and (100), which have double bonds in these positions, and so provide strong evidence for path c in Scheme 26. Consistent with the previous results from incorporation of [1, 2-¹³C₂]acetate, there is no randomisation of labelling between C-14 and C-15, so the stereochemical integrity of the *gem*-dimethyl group is retained throughout the biosynthetic pathway which perhaps suggests that the alkene (96) is formed by a concerted elimination from the ε-lactone system (95).

Austin (89) is clearly biosynthetically related to andibenin B and is formed from the same key intermediate (86). An interesting possibility was that the tetracyclic intermediate (91) involved in austin biosynthesis (Scheme 24) is not formed directly from cyclisation of (86) but is formed via the same bicyclofarnesyl intermediate (87) involved in andibenin biosynthesis (Scheme 22). To test this, [6-¹³C, 6-²H₃]mevalonate was fed to 5-day old cultures and the resultant enriched austin isolated and its ¹H, ²H noise decoupled ²H nmr spectrum determined. This showed clear isotopically shifted signals corresponding to the incorporation of two and mainly three deuteriums into the 12-, 13- and 14-methyl groups. The result for the 12-methyl group excludes the possibility of the involvement of (87) in austin biosynthesis.

The labelling studies described above provide definitive evidence for the mixed polyketide-terpenoid biogenesis of the andibenins, andilesins, anditomins, austin and terretonin. The formation of the bicyclo[2.2.2]octane system in the first two classes of metabolite provides a rare example of a biosynthetic Diels-Alder reaction. The biosynthetic relationship of austin and andibenin was supported by the isolation of austin from another mutant strain of *A. varicolor* [81]. Further metabolites related to austin have been isolated from *Emericella dentata* [82] and *Penicillium diversum* [81]. Other complex metabolites which are almost certainly further products of the meroterpenoid pathway are fumigatonin (102) and paraherquonin (103) which have been isolated from *Aspergillus*

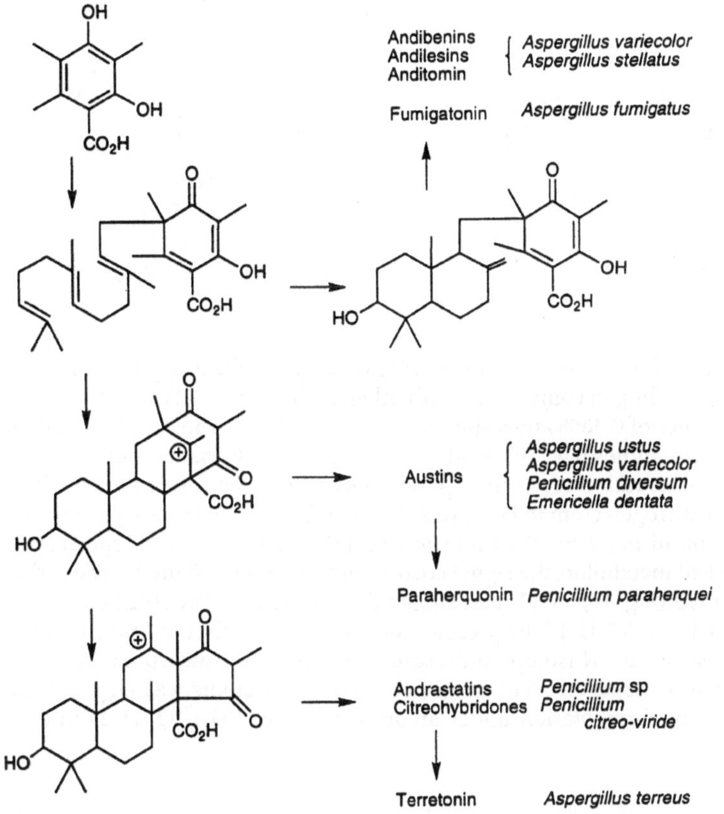

Scheme 27

fumigatus [83] and *Penicillium paraherquei* [84] respectively. More recently the citreohybridones, e.g. (104), and the closely related andrastatins have been isolated from *Penicillium citreo-viride* [85] and *Penicillium* sp. OF-3929 [86] respectively, providing yet further members of the family. Labelling studies [87] with ^{13}C-labelled acetates and [*carboxy*, 2-^{13}C$_2$]-3,5-dimethylorsellinate into citreohybridone (104) are entirely consistent with these proposals. These latter metabolites are clearly related to terretonin (90) and the likely biogenetic relationships amongst these structurally varied metabolites are summarised in Scheme 27.

7
3-Nitropropanoic Acid: ^{15}N Labelling Studies

3-Nitropropanoic acid (109) is a toxic metabolite produced in higher plants and by several fungi. It has been shown to be produced in *Penicillium atrovenetum* from L-aspartate (105) (Scheme 28). A series of papers has been published on the

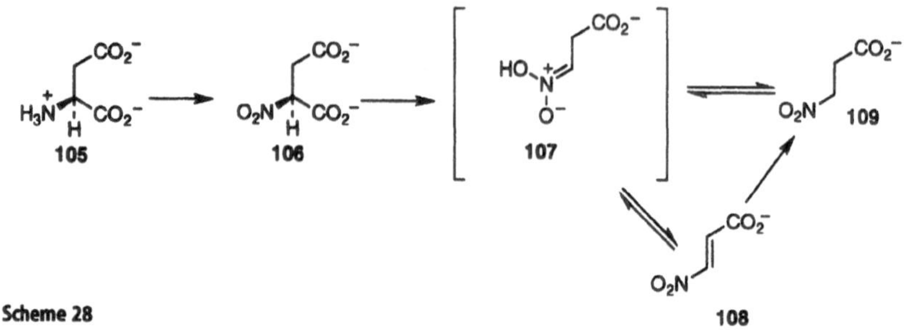

Scheme 28

details of this conversion which provide some excellent applications of the use of ^{15}N labelling in conjunction with other stable isotopes. Since ^{15}N has a natural abundance of 0.36% and a spin of 1/2, ^{15}N NMR is possible. Although most biosynthetic studies have detected incorporation of ^{15}N via its coupling to ^{13}C, the direct observation of ^{15}N incorporation by NMR spectroscopy is possible.

The nitrogen atom in (109) was shown to be derived from aspartate by incorporation of DL-[2-^{13}C,^{15}N]aspartic acid [88]. In the ^{13}C NMR spectrum of the enriched metabolite, the signal corresponding to the C-3 methylene carbon was observed (Fig. 14) as a broad singlet due to ^{13}C with ^{14}N attached and a sharp doublet (J_{CN} 8.7 Hz). The presence of the doublet, offset to lower frequency by the associated ^{15}N isotope shift, confirmed the intact incorporation of the C-N bond from aspartate. The origin of the oxygen atoms was elegantly demonstrated by fermentation under an atmosphere of $^{16}O_2$:$^{18}O_2$ (1:1) in a defined

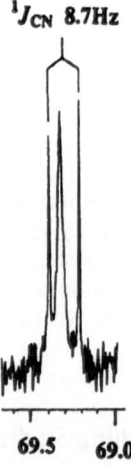

$^{1}J_{CN}$ 8.7Hz

69.5   69.0

Fig. 14. The C-3 methylene region of the 50 MHz p.n.d. ^{13}C NMR spectrum of 3-nitropropanoic acid (109) enriched from feeding DL-[2-^{13}C,^{15}N]aspartic acid to *Penicillium atrovenetum*. (Figure kindly supplied by R. L. Baxter)

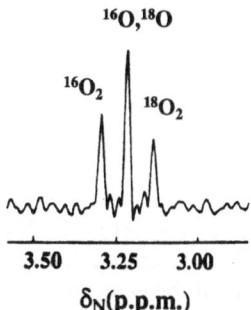

Fig. 15. 36.5 MHz p.n.d. ^{15}N NMR spectrum of 3-nitropropanoic acid biosynthetically enriched by growing *Penicillium atrovenetum* on [15]NH$_4$Cl under an [18]O$_2$ atmosphere. The observed [18]O isotope induced shift is 0.08 ppm per oxygen atom. (Figure kindly supplied by R.L. Baxter)

medium in which [15]NH$_4$Cl was the sole nitrogen source [89]. The ^{1}H NMR spectrum of the isolated 3-nitropropanoic acid showed 2 and 3-bond [15]N-^{1}H couplings which on integration confirmed a > 95% [15]N abundance. The ^{15}N DEPT NMR spectrum (Fig. 15) showed three signals corresponding to [15]N[16]O$_2$, [15]N[16]O[18]O and [15]N[18]O$_2$ species, proving that both oxygen atoms of the nitro group are derived from molecular oxygen. [^{15}N]-Nitrosuccinate (**106**) was shown to be an intermediate by ^{15}N NMR analysis of the metabolite isolated after feeding the substrate as its diethyl ester [90]. Assignment of the stereochemistry of the intermediate nitrosuccinate was ascertained indirectly by examining the incorporation of deuterium from L-[2, 3, 3-[2]H$_3$]aspartate into (**109**). The resulting ^{2}H NMR spectrum shows that ^{2}H is retained at both the 2- and 3-positions, indicating that the stereochemistry of the L-aspartate is retained in the oxidation of the amino acid to nitrosuccinate and so makes inversion unlikely. As a result of these and further experiments with doubly ^{13}C, ^{2}H labelled aspartates and 3-nitropropanoic acid which revealed a futile cycle involving 3-nitroacrylate (**108**), the pathway summarised in Scheme 28 was proposed [91]. Oxidation of L-aspartate affords (*S*)-nitrosuccinate which on decarboxylation produces and *aci*-nitro-intermediate (**107**) as the kinetic product which tautomerises to 3-nitropropanoic acid. When nitropropanoate dehydrogenase is present, (**107**) or (**109**) may be intercepted and oxidised to 3-nitroacrylate. It was suggested that this cycle may serve as an additional mechanism for regulating NADP/NADPH levels in the cell.

8
Isotopic Labelling of Biosynthetic Proteins

Stable isotope labelling is also proving to have an important role in studies on the enzymology of polyketide biosynthesis. The acyl carrier protein (ACP) components of polyketide synthases (PKSs) are believed to play a central role in the control of the assembly and stabilisation of polyketide intermediates, especially of the highly oxygenated intermediates necessarily involved in biosynthesis of

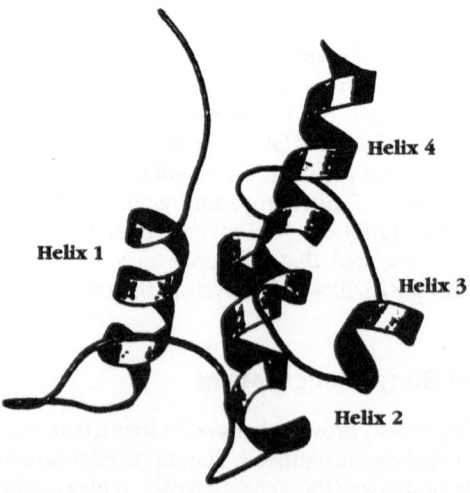

110   111

the polycyclic aromatic antibiotics, e.g. actinorhodin (110) and oxytetracycline (111), produced by Type II PKSs in Streptomycetes [7]. A number of these ACPs have been isolated after cloning of the gene and overexpression in *E. coli*. [92]. The full 3D structure of actinorhodin ACP (Fig. 16) has been elucidated by application of 2D NMR methods [93, 94]. The ACP contains 14 leucine residues out of a total of 82 amino acid residues and a crucial part of the refinement of the structure was the biosynthetic labelling of the protein by expression in *E. coli* grown on a minimal medium supplemented with leucine labelled stereospecifically in the diastereotopic methyl groups.

The specifically labelled leucine (114) was prepared [95] as shown in Scheme 29 by a chemoenzymatic synthesis in which the stereogenic centre at C-4 was established by treatment of the enolate of the propanoyl oxazolidinone (112) with CD_3I. Cleavage of the oxazolidinone followed by two carbon homologations via a Grignard reaction gave the α-ketoester (113) selectively labelled with deuterium. Saponification of the ester and reductive amination of the ketone catalysed by leucine dehydrogenase gave $(2S,4R)$-[5, 5, 5-$^2H_3$]leucine. Related chemoenzymatic routes can be used to prepare readily a wide range of isotopically labelled amino acids for protein structural studies and metabolic studies [96].

Fig. 16. Molscript diagram showing the restrained minimised structure of actinorhodin *apo* ACP

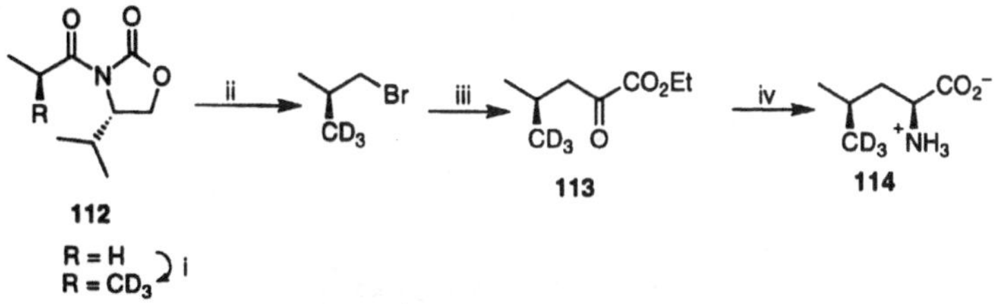

Reagents: i) LDA, CD₃I; ii) LiAlH₄, THF then Ph₃P, Br₂, PhNO₂; iii) (CO₂Et)₂, Mg, Et₂O, THF; iv) NaOH then leucine dehydrogenase.

Scheme 29

Part of the $^1H,^1H$ COSY NMR spectra of the methyl region of the unlabelled and labelled ACP are shown in Fig. 17. This shows the absence of one set of leucine methyls in the labelled protein. The effect is most clearly seen by looking at the signals assigned to leucine 45. These appear at 0.28 and 0.38 ppm in the unlabelled spectrum. In the labelled spectrum the signal at 0.38 ppm has been completely eliminated, allowing the assignment of the remaining signal to the pro-S methyl of leucine 45.

For in vitro studies of polyketide assembly using the purified components of the Type II PKS, it is essential to have the ACPs in their active *holo*-form which can then be acylated chemically or enzymatically (Scheme 30). The inactive apo-ACPs are converted to the active holo-form by addition of phosphopantetheine from coenzyme A to a conserved serine residue. This addition is mediated in each organism by an ACP-holo synthase. In *E. coli* there is an ACP-holo synthase which is part of its endogenous fatty acid biosynthetic machinery but the substrate specificity of this enzyme for heterologous ACPs has been shown

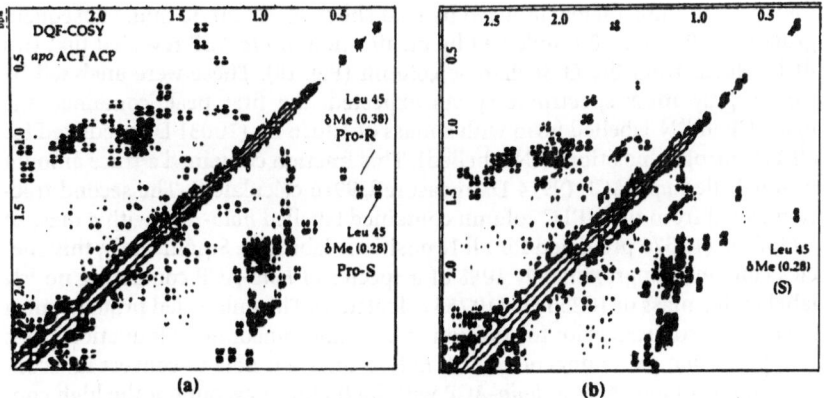

Fig. 17 a, b. 500 MHz DQF-COSY NMR spectra of actinorhodin ACP; a unlabelled; b labelled from incorporation of $(2S,4R)$-[5, 5, 5-$^2H_3$]leucine

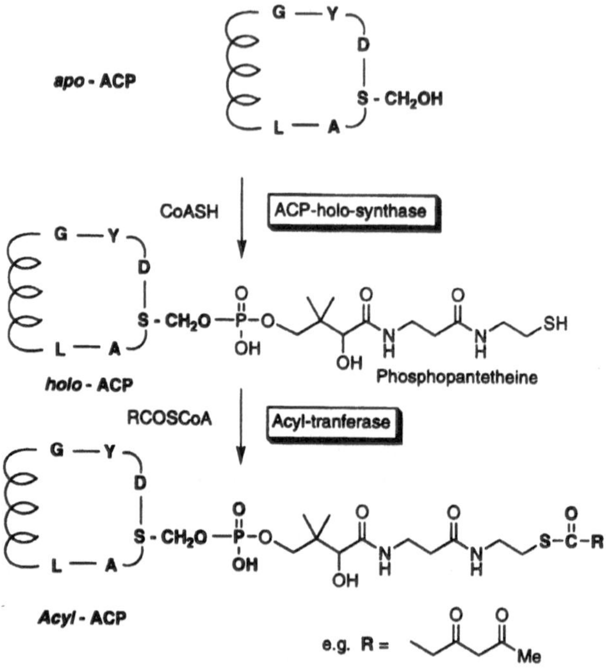

Scheme 30

[92, 97] to vary and is generally low for type II PKS ACPs. An interesting observation was made during the preparation of ^{15}N-labelled oxytetracyline ACP which was required for 3D NMR studies [98]. Thus *E. coli* containing the *otc* ACP gene on a heat-inducible expression system was grown at 30 °C to mid log phase. The cells were then washed and transferred to minimal medium containing ^{15}NH${}_4$Cl and immediately induced by heat shock to 42° for 30 min. Subsequent growth at 30 °C for 18 h followed by purification of *otc* ACP revealed that two ACPs eluted from the Q-Sepharose column (Fig. 18). These were analysed by electrospray mass spectrometry. As expected, the first peak contained the *apo*-ACP in ^{15}N-labelled form with a mass of 10026 Da (10031 Da predicted for all 115 nitrogen positions ^{15}N labelled). This fraction contained a trace amount of unlabelled *apo*-ACP (9914 Da measured, 9916 calculated). The second fraction eluted from the HPLC column contained labelled *holo*-ACP with a mass of 10367 Da (10376 predicted for all N positions labelled). Significantly, this species contained approximately 10% of a species of *holo*-ACP containing no ^{15}N label with a mass of 10256 Da (10256 calculated). The unlabelled proteins must have been produced prior to addition of label and concomitant induction. Thus it appears that the endogenous ACP *holo*-synthase is able to convert low concentrations of *apo*-ACP to *holo*-ACP with high efficiency, but not the high concentrations produced after induction. This may provide evidence for substrate inhibition of the enzyme.

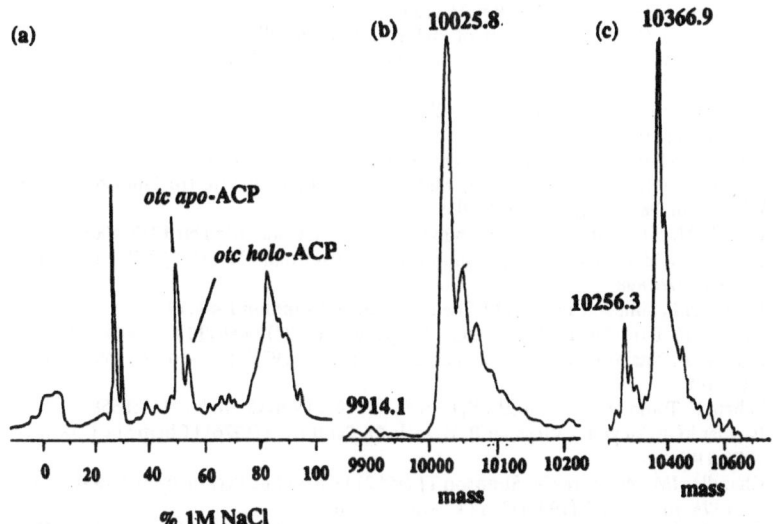

Fig. 18 a–c. Purification and ESMS analysis of oxytetracycline acyl carrier proteins after growth and induction in minimal medium containing $^{15}NH_4Cl$: **a** Q-Sepharose FPLC trace of *E. coli* protein fractions containing ACP; **b** ESMS of ^{15}N-oxytetracycline *apo*-ACP: note distinctive envelope of Na$^+$ adducts; **c** ESMS of ^{15}N-oxytetracycline *holo*-ACP, again showing Na$^+$ adducts, as well as a significant proportion (ca. 10%) of unlabelled *holo*-ACP

9
Conclusions

There can be little doubt that the application of stable isotope labelling methodologies has made a major impact on the development of biosynthetic and metabolic studies in a wide range of systems. As biosynthetic work becomes increasingly focused on studies at the enzymatic level, there will be continued scope for the further development of these applications particularly making use of sensitive mass spectral detection methods rather than the NMR methods which have dominated whole cell studies. We look forward to the results of these studies with anticipation.

10
References

1. Simpson TJ (1975) Chem Soc Rev 4:497
2. Garson MJ, Staunton J (1979) Chem Soc Rev 8:539
3. Abell C (1986) The use of NMR spectroscopy to follow deuterium in studies of fungal metabolism. In:Linskens HF, Jackson JF (eds) Modern methods of plant analysis vol 2. Springer, Berlin Heidelberg New York, p 60
4. Simpson TJ (1986) ^{13}C NMR in metabolic studies. In: Linskens HF, Jackson JF (eds) Modern methods of plant analysis vol 2. Springer, Berlin Heidelberg New York, p 1

5. Simpson TJ (1987) Chem Soc Rev 16 : 123
6. Simpson TJ (1991) Application of stable isotope labelling and multinuclear NMR to biosynthetic studies. In:Buncel E, Jones JR (eds) Isotopes in the physical and medical sciences. Elsevier, Amsterdam, p 431
7. Simpson TJ (1995) Chem Ind 1995:407
8. Simpson TJ (1979) J Chem Soc Perkin Trans 1 1979:1233
9. Overton KH, Picken DJ (1976) J Chem Soc Chem Comm 1976:105
10. (a) Sadler IH, Simpson TJ (1992) Magn Reson Chem 30:518; (b) Simpson TJ, (1994) J Chem Soc Perkin Trans 1 1994:3055
11. Kaneda M, Takahashi R, Iitaka Y, Shubata S (1972) Tetrahedron Lett 1972:4008
12. Birch AJ, Baldas J, Hlubucek JR, Simpson TJ, Westerman PW (1976) J Chem Soc Perkin Trans 1 1976:898
13. Bardshiri E, Simpson TJ (1981) J Chem Soc Chem Commun 1981:195
14. Ahmed SA, Bardshiri E, McIntyre CR, Simpson TJ (1992) Austral J Chem 45:249
15. Arnone A, Nasini G, Merlini L, Ragg E, Assante G (1993) J Chem Soc Perkin Trans 1 1993:145
16. Tabata N, Tomoda H, Matsuzaki K, Omura S (1993) J Am Chem Soc 115:8558
17. Pachler KGR, Steyn PS, Vleggaar R, Wessels PL, Scott De B (1976) J Chem Soc Perkin Trans 1976:1182
18. Chandler IM, McIntyre CR, Simpson TJ (1992) J Chem Soc Perkin Trans 1 1992:2285
19. Ayer WA, Jiminez LD (1994) Can J Chem 72:2326
20. Chandler IM, McIntyre CR, Simpson TJ (1992) J Chem Soc Perkin Trans 1 1992:2271
21. Gudgeon JA, Holker JSE, Simpson TJ (1974) J Chem Soc Chem Commun 1974:636
22. Gudgeon JA, Holker JSE, Simpson TJ, Young K (1979) Bioorganic Chemistry 8:311
23. Holker JSE, O'Brien E, Moore RN, Vederas JC (1983) J Chem Soc Perkin Trans 1 1983:192
24. Simpson TJ, Holker JSE (1975) Tetrahedron Lett 1975:4693
25. Nakajima H, Fukuyama K, Fujimoto H, Baba T, Hamasaki T (1994) J Chem Soc Perkin Trans 1 1994:1865
26. MacMillan J, Lunnon MW (1976) J Chem Soc Perkin Trans 1 1976:584
27. Simpson TJ, Stevenson GI (1985) J Chem Soc Chem Commun 1985:1822
28. Simpson TJ, Stenzel DJ (1982) J Chem Soc Chem Commun 1982:1074
29. Leeper FJ, Staunton J (1982) J Chem Soc Chem Commun 1982:911
30. O'Neill JA, Simpson TJ, Willis CL (1993) J Chem Soc Chem Commun 1993:738
31. O'Neill JA, Simpson TJ, Willis CL (1994) J Chem Soc Perkin Trans 1 (1994):2493
32. O'Neill JA, Lindell SD, Simpson TJ, Willis CL (1996) J Chem Soc Perkin Trans 1 1996:632
33. Ahmed SA, Simpson TJ, Staunton J, Sutkowski AC, Trimble LA, Vederas JC (1985) J Chem Soc Chem Commun 1985:1685
34. Staunton J, Sutkowski AC (1991) J Chem Soc Chem Commun 1991:1106; Laue ED, Pollard KOB, Skilling J, Staunton J, Sutkowski AC (1987) J Mag Res 72:493
35. Staunton J, Sutkowski AC (1991) J Chem Soc Chem Commun 1991:1108
36. Staunton J, Sutkowski AC (1991) J Chem Soc Chem Commun 1991:1110
37. Jacobs A, Staunton J, Sutkowski AC (1991) J Chem Soc Chem Commun 1991:1113
38. Jacobs A, Hill A, Staunton J (1995) J Chem Soc Chem Commun 1995:859
39. Hill A, Staunton J (1995) J Chem Soc Chem Commun 1995:861
40. Jacobs A, Staunton J (1995) J Chem Soc Chem Commun 1995:863
41. Cane D, Yang CC (1987) J Am Chem Soc 109:1255
42. Hutchinson CR, Yue S, Duncan JS, Yamamoto Y (1987) J Am Chem Soc 109:1253
43. Yoshizawa Y, Li Z, Reese PB, Vederas, JC (1990) J Am Chem Soc 112:212
44. Li Z, Martin FM, Vederas J (1992) J Am Chem Soc 114:1531
45. Scott FE, Simpson TJ, Trimble LA, Vederas JC (1984) J Chem Soc Chem Commun 1984:756
46. McNicholas C, Simpson TJ, Willett NJ (1996) Tetrahedron Lett 37:8053
47. Rawlings BJ, Reese PB, Ramer SE, Vederas JC (1989) J Am Chem Soc 111:3382
48. Arai K, Rawlings BJ, Yoshizawa Y, Vederas JC (1989) J Am Chem Soc 111:3391
49. Townsend CA, Brobst SW, Ramer SE, Vederas JC (1988) J Am Chem Soc 110:318
50. Brobst SW, Townsend CA (1994) Can J Chem 72:200

51. Watanabe CMH, Wilson D, Linz JE, Townsend CA (1996) Chemistry and Biology 3:463
52. Abraham EP, Pang, CP, White RL, Crout DHG, Mutstorf L, Morgan PJ, Derome AE (1983) J Chem Soc Chem Commun 1983:723
53. Evans JNS, Fagerness PE, Mackenzie NE, Scott AI (1984) J Am Chem Soc 106:5738
54. Hayes MA (1991) PhD thesis, University of Bristol
55. Abell C, Staunton J (1984) J Chem Soc Chem Commun 1984:1005
56. Jordan PM, Spencer JB (1990) J Chem Soc Chem Commun 1990:238
57. Jordan PM, Spencer JB (1990) J Chem Soc Chem Commun 1990:1704
58. Spencer JB, Jordan PM (1992) J Chem Soc Chem Commun 1992:646
59. Woo E-R, Fujii I, Ehizuka E, Sankawa U, Kamaguchi A, Huang S, Beale JM, Shibuya M, Mocek U, Floss HG (1989) J Am Chem Soc 111:5498
60. Thiericke R, Rohr J (1993) Nat Prod Rep 10:265
61. Harper DB, O'Hagan D (1994) Nat Prod Rep 11:123
62. Jones CA, Sidebottom PJ, Cannell RJP, Noble D, Rudd, BAM (1992) J Antibiotics 45:1492
63. Cannell RJP, Dawson MJ, Hale RS, Hall RM, Noble D, Lynn S, Taylor NL (1993) J Antibiotics 46:1381
64. Less SL, Handa S, Millburn K, Leadlay PF, Dutton CJ, Staunton J (1996) Tetrahedron Lett 37:3515
65. Less SL, Leadlay PF, Dutton CJD, Staunton J (1996) Tetrahedron Lett 37:3519
66. Less SL, Handa S, Leadlay PF, Dutton CJD, Staunton J (1996) Tetrahedron Lett 37:3511
67. McKeown DSJ, McNicholas C, Simpson TJ, Willett NJ (1996) J Chem Soc Chem Commun 1996:301
68. Simpson TJ (1979) J Chem Soc Perkin Trans 1 1979:2118
69. Simpson TJ, Walkinshaw MD (1981) J Chem Soc Chem Commun 1981:914
70. Holker JSE, Simpson TJ (1978) J Chem Soc Chem Commun 1978:626
71. Chexal KK, Springer JP, Clardy J, Cole RJ, Kirksey TW, Dorner JW, Cutler HG, Strawter WJ (1976) J Am Chem Soc 98:6748
72. Springer JW, Corner JW, Cole RJ, Cox RH (1979) J Org Chem 44:4852
73. Bartlett AJ, Holker JSE, O'Brien E, Simpson TJ (1983) J Chem Soc Perkin Trans 1 1983:667
74. Bartlett AJ, Holker JSE, O'Brien E, Simpson TJ (1981) J Chem Soc Chem Commun 1981:1198
75. Ahmed S, Scott FE, Simpson TJ, Trimble LA, Vederas JC (1989) J Chem Soc Perkin Trans 1 1989:807
76. McIntyre CR, Scott FE, Simpson TJ, Trimble LA, Vederas JC, (1989) Tetrahedron 45:2307
77. McIntyre CR, Simpson TJ, Moore RN, Trimble LA, Vederas JC (1984) J Chem Soc Chem Commun 1984:1498
78. McIntyre CR, Scott FE, Simpson TJ, Trimble LA, Vederas JC (1986) J Chem Soc Chem Commun 1986:501
79. Scott FE, Simpson TJ, Trimble LA, Vederas JC (1986) J Chem Soc Chem Commun 1986:214
80. Simpson TJ, Ahmed SA, McIntyre CR, Scott FE, Sadler IH (1997) Tetrahedron 53:4013
81. Simpson TJ, Stenzel DJ, Bartlett AJ, O'Brien E, Holker JSE (1982) J Chem Soc Perkin Trans 1 1982:2687
82. Maebayashi Y, Okuyama M, Yamakazi M, Katsube Y (1982) Chem Pharm Bull 30:1911
83. Okuyuma E, Yamasaki M, Katsube Y (1984) Tetrahedron Lett 25:3233
84. Okuyama M, Yamasaki M, Kobayashi K, Sakuvai T (1983) 24:3113
85. Kosemura S, Matsunaga K, Yamamuro S (1991) Chemistry Lett 1991:811
86. Shiomi K, Uchida R, Inokoshi J, Tanaka H, Iwai T, Omura S (1996) Tetrahedron Lett 37:1265
87. Kosemura S, Miyata H, Yamamura S, Albone K, Simpson TJ (1994) J Chem Soc Perkin Trans 1 1994:135
88. Baxter RL, Abbot EM, Greenwood SL, MacFarlane IJ (1985) J Chem Soc Chem Commun 1985:564
89. Baxter RL, Greenwood SL (1986) J Chem Soc Chem Commun 1986:175
90. Baxter RL, Hanley AB, Chan HW-S (1988) J Chem Soc Chem Commun 1988:757

91. Baxter RL, Hanley AB, Chan HW-S, Greenwood SL, Abbot EM, MacFarlane IJ, Milne K (1992) J Chem Soc Perkin Trans 1 1992:2495
92. Crosby J, Sherman DH, Bibb M, Simpson TJ, Hopwood DA, (1995) Biochim Biophys Acta 1251:32
93. Crump MP, Crosby J, Dempsey CE, Murray M, Simpson TJ, Hopwood DA (1996) FEBS Letters 391:302
94. Crump MP, Crosby C, Dempsey CE, Parkinson JA, Murray M, Hopwood, DA, Simpson TJ (1997) Biochemistry 36:6000
95. Kelly NM, Reid RG, Willis CL, Winton P (1995) Tetrahedron Lett 36:8315
96. Simpson TJ, Willis CL (1996) Enzymatic and chemical methods for the enantioselective synthesis of isotopically labelled amino acids and metabolic intermediates. Proceedings of the Chiral Europe '96 Symposium, Strasbourg, p 49
97. Gehring AM, Lambalot RM, Vogel KW, Drueckammer DG, Walsh CT (1997) Chemistry & Biology 4:17
98. Cox RJ, Hitchman TS, Byrom KJ, Findlow ISC, Tanner JA, Crosby J, Simpson TJ (1997) FEBS Letters 405:267

The Biosynthesis of Aliphatic Polyketides

James Staunton[1*] · Barrie Wilkinson[2]

[1] University Chemical Laboratory, Lensfield Road, Cambridge, CB2 1EW, Great Britain.
   E-mail: js24@cus.cam.ac.uk
[2] Bioprocessing Research Unit, GlaxoWellcome Research and Development,
   Medicines Research Centre, Gunnels Wood Road, Stevenage, Hertfordshire, SG1 2NY,
   Great Britain.

Aliphatic polyketides are a large family of natural products which exhibit an impressive range of biological activities. They are synthesised by a common pathway in which units derived from acetate, propionate and butyrate are condensed onto a growing chain, which is initiated from a range of structurally varied carboxylic acid starter units, in a process closely resembling fatty acid biosynthesis. The intermediates remain bound to the polyketide synthase (PKS) throughout multiple rounds of chain extension, and to a variable degree, reduction of the newly formed β-keto functionality. It is the manner in which each different PKS controls the number and type of extender units added, and the extent and stereochemistry of reduction at each cycle which gives rise to the diverse structural variation seen in this family of natural products. Furthermore, the aglycone product of the PKS may be modified by post-PKS enzymes, including glycosidases, methyltransferases, acylases and oxidative enzymes, to produce still greater diversity. In this article the development of the field will be exemplified by a detailed discussion of the archetypal example of erythromycin A biosynthesis. This will then be followed by further discussion of other relevant areas of research in this field.

Keywords: Aliphatic polyketides; Biosynthesis; Polyketide Synthase; Genetic Engineering; Erythromycin A; Rapamycin.

* Corresponding author.

Topics in Current Chemistry, Vol. 195
© Springer Verlag Berlin Heidelberg 1998

List of Abbreviations and Symbols

ACP	Acyl Carrier Protein
AHBA	3-Amino-5-Hydroxybenzoic Acid
AT	Acyl Transferase
ATP	Adenosine 5-Triphosphate
bp	base pairs
CHC	Cyclohexane Carboxylic Acid
CL	Co-A Ligase
DEBS	6-Deoxyerythronolide B Synthase
DH	Dehydratase
DHAP	3-Deoxy-D-*arabino*-heptulosonic Acid-Phosphate
DHCHC	*trans*-Dihydroxy Cyclohexane Carboxylic Acid
ER	Enoyl Reductase
FAS	Fatty Acid Synthase
kb	kilobases
KR	β-Keto Reductase
KS	β-Keto Synthase
NAC	*N*-Acetylcysteamine
NADPH	β-Nicotinamide Adenine Dinucleotide Phosphate, Reduced Form
ORF	Open Reading Frame
PKS	Polyketide Synthase
PMSF	Phenylmethylsulfonyl Fluoride
SAM	*S*-Adenosyl Methionine
TE	Thioesterase

1
Introduction to Polyketide Structures

The common feature of polyketide natural products is that they are biosynthesised by repeated addition of small building blocks, usually comprising two
or three carbons, to form a linear chain. The resulting chain can be a polyketone
1 or it can have an array of functional groups formally derived by reductive
modification of most or all of the keto groups of a polyketone. Typical modified
residues are shown in 2–4. The length of chain may vary greatly, although the
majority of polyketides incorporate from four to ten building blocks. It is these
heavily reduced aliphatic polyketide chains and molecules derived from them
which form the subject of this article. Only such polyketides of terrestrial, and
not marine origin will be considered.

The Archetypal Poly-β-keto Chain   Structural Variations Produced by Reduction of Keto Groups

The initial polyketide chain may survive unaltered, but almost always it is
further elaborated by additional biosynthetic operations to give a structurally
more complex product. Examples of these elaboration processes are oxidative
changes, such as hydroxylation, epoxidation or oxidative cleavage, or addition of
extra structural residues by esterification, alkylation or glycosylation. If the
elaboration steps involve a cleavage or rearrangement of the primary polyketide
chain, a structure may result which is radically different from the archetypal
polyketide skeletal type represented by 5. Even so, the natural product is cate-
gorised as a polyketide by virtue of its biosynthetic origin.

The combination of variety in the processes of chain assembly and elaboration
makes for great diversity of structure; indeed, by taking into account all the possible biosynthetic variations it can be calculated that potentially there are tens,
possibly hundreds, of millions of different structures waiting to be discovered. In
fact, after decades of searching through metabolites of both marine and terrestrial
organisms, the number of known compounds runs only into tens of thousands.
New structures are being discovered at what seems an impressive rate until the
potential for diversity is taken into account. The present limit of known diversity
is further evident when rules of structural relationship such as "Celmer's" rules are
considered [1]. The known compounds are clustered into families of related structures which may have a common evolutionary origin. The diversity map which
then emerges is one of a few tiny islands of structure distributed over a blank
ocean in which potential structures are absent. One of the challenges in this field
is to understand the genetics of the biosynthetic processes with a view to expanding
the range of structural diversity by genetic engineering. This goal, which seemed a
fanciful idea as recently as 1990, had already been achieved five years later.

Many natural polyketides have been investigated with differing degrees of success and progress. Not surprisingly, most interest has focused on commercially important compounds which are useful in medicine or agriculture, and most progress has been made on one such compound, the antibiotic erythromycin. Accordingly, in this account erythromycin will be covered first and in great detail to show how far the field has developed. Other important biosynthetic pathways which are beginning to be understood at the genetic level will then be presented for comparison and to show the exciting potential for the discovery of new compounds offered by this field.

2
The Erythromycins

2.1
Structures of the Erythromycins

The parent compound in the erythromycin family, erythromycin A 5, is typical of a large family of macrolides which are characterised by rings containing 12, 14 or 16 atoms [2]. Members of the family show an intriguing common structural and stereochemical relationship which was first noted in Celmer's Rules [1]. Erythromycin A was isolated in 1952 from *Saccharopolyspora erythraea* [3]. It is used in clinical medicine against infections caused by Gram-positive bacteria, and it is the main treatment for many pulmonary infections such as Legionnaire's disease. The structure was elucidated by classical methods [4] and confirmed by X–ray crystallography [5]. Other members of the family occur as biosynthetic intermediates and so will be presented in that context later.

2.2
Overview of the Erythromycin Biosynthetic Pathway

The biosynthesis of erythromycin can be divided into two phases (Scheme 1). In the first constructive phase of the pathway a set of key enzymes, collectively known as the polyketide synthase (PKS), assembles the typical polyketide chain by sequential condensation of one unit of propionyl-CoA and six units of methylmalonyl-CoA 6. The initially formed chain is cyclised to give the first macrocyclic lactone (macrolide) intermediate 6-deoxyerythronolide B 7 [6, 7]. In the second phase 6-deoxyerythronolide B is elaborated by a series of "tailoring" enzymes which carry out regiospecific hydroxylations, glycosylations and a methylation (of an added sugar residue) to give finally erythromycin A. The core polyketide structure is generated by the PKS in phase one, but the later steps of phase two are essential to produce active antibiotics.

2.3
The Elaboration Steps of the Pathway

In a search for new antibiotics or to improve the yields of the existing compounds, the producing organism was subjected to intensive mutation studies in

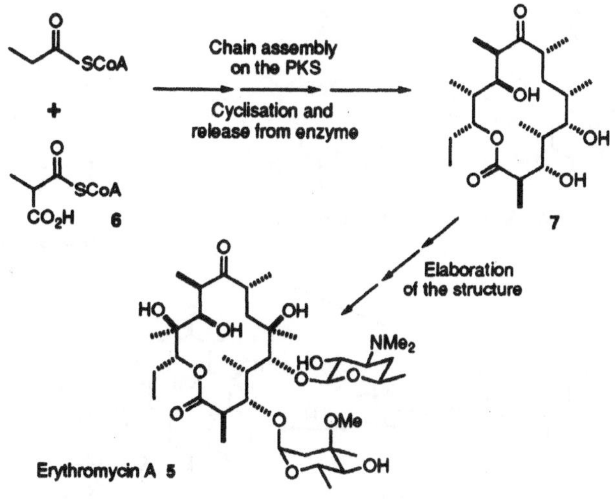

Scheme 1

the hope of modifying the biosynthetic machinery. Most of the information concerning the late stages of the biosynthetic pathway came adventitiously as a spin-off from these studies through the chance generation of blocked mutants. When a late step on the pathway is blocked by lack of the requisite enzyme the previous intermediate may accumulate in sufficient quantities for it to be isolated and identified. Mutants blocked in each step in the pathway from 6-deoxyerythronolide B 7 to erythromycin A 5 have been produced, allowing the late intermediates to be identified [8] (Scheme 2).

The first step is hydroxylation at C-6 to give erythronolide B 8 [9]. Its intermediacy was confirmed by incorporation of radioactive 7 [10] and isotope dilution experiments [11]. Subsequently, L-mycarose is attached to the C-3 hydroxyl group of 8 by a TDP–mycarose glycosyltransferase [12, 13], and the amino sugar D-desosamine is then added to the C-5 hydroxyl of 9 by the action of the enzyme TDP-desosamine glycosyltransferase [13, 14]. The resulting intermediate, erythromycin D 10, is the first to show antibacterial activity and occurs at a branch in the biosynthetic pathway. Either C-12 hydroxylation takes place with retention of configuration [15], to produce erythromycin C 11, or O–methylation of the C-3″ hydroxyl of the mycarose moiety with SAM, catalysed by an O–methyltransferase, produces erythromycin B 12. Finally erythromycin A 5 can be generated either by O–methylation of 11 catalysed by an O–methyltransferase and SAM, or by C-12 hydroxylation of 12. A single O–methyltransferase operates on both branches of the pathway [16]. Similarly a single hydroxylase has been implicated [17]. Since this has a 1200–1900-fold preference for erythromycin C over B it is considered that the pathway via erythromycin B is the minor one.

The biosynthetic pathways leading to the sugar residues have not been established in detail, but a speculative pathway has been proposed based on genetic analysis [18, 19].

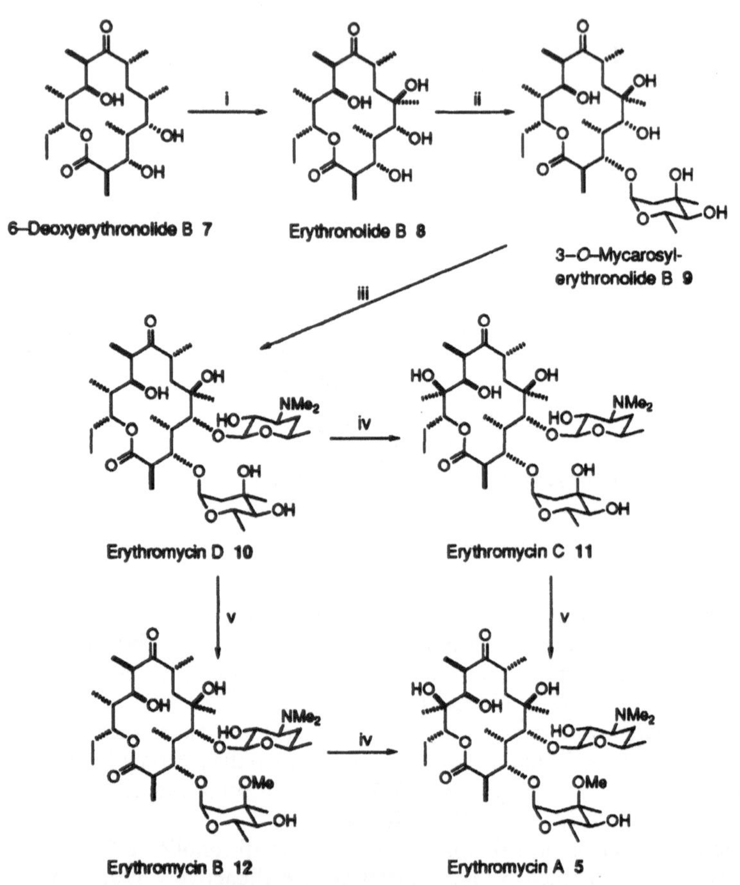

6–Deoxyerythronolide B **7**

Erythronolide B **8**

3–O–Mycarosyl-erythronolide B **9**

Erythromycin D **10**

Erythromycin C **11**

Erythromycin B **12**

Erythromycin A **5**

i, C-6 erythronolide hydroxylase; ii, TDP–mycarose glycosyltransferase; iii, TDP–desosamine glycosyltransferase; iv, C-12 hydroxylase; v, O-methyltransferase

Scheme 2

2.4
Biosynthesis of the Macrolide Core – Classical Precursor Studies

Corcoran and co-workers first fed labelled precursors to *S. erythraea* and detected their incorporation into 6–deoxyerythronolide B **7** (Scheme 3) [20]. These results gave the first evidence that the macrolide core is derived from one propionyl-CoA and six methylmalonyl CoA units. Subsequently Cane and co-workers [21] fed [^{18}O]propionate and demonstrated that all the oxygens attached to carboxyl-derived carbons in the macrolide core were retained from the propionate precursor and were not derived from molecular oxygen or water. These results are consistent with reduction of each keto group of the putative

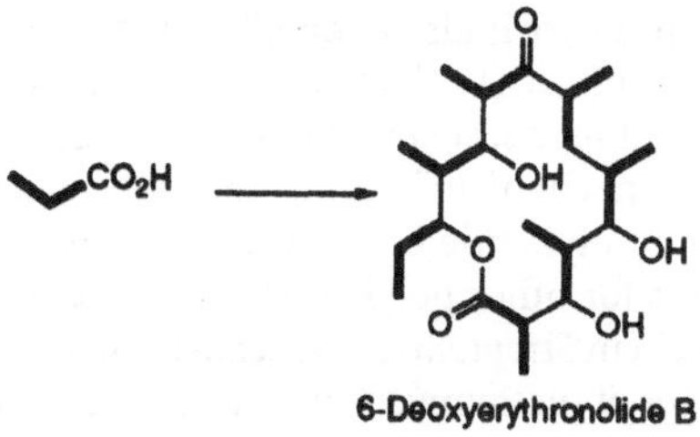

Scheme 3   6-Deoxyerythronolide B **7**

polyketone chain to the required oxidation state and stereochemistry, although the timing of the reduction steps in relation to the chain extension processes remained to be established.

In studies using blocked mutants, no intermediates preceding 6-deo-xyerythronolide B were isolated, implying that these intermediates are either too unstable to accumulate or enzyme-bound [8]. Subsequently in an important pioneering experiment, Cane and Yang investigated a possible diketide inter-mediate in which the keto group has been reduced to hydroxyl [22]. The free acid showed random incorporation into erythromycin B **12** indicating that the diketide had been degraded to propionate before being used in the biosynthesis. However, the *N*–acetylcysteamine (NAC) thioester analogue **13** was incorpo-rated intact with retention of the coupling between the two simultaneously labelled sites of ^{13}C enrichment (Scheme 4). To demonstrate that **13** was not oxi-dised to the β-keto thioester before incorporation an analogous deuterium labelled diketide **14** was fed and was again incorporated intact, thus eliminating this possibility [23]. The *N*–acetylcysteamine (NAC) thioester was employed because it exhibits a high degree of structural homology with the thiol terminus of coenzyme A and of the 4′-phosphopantetheine group of the active acyl carrier protein. This experiment, with a similar contemporary pioneering contribution from Hutchinson et al. [24], opened up a highly productive phase of research into the biosynthesis of other polyketide metabolites (see later). From these results it can be concluded that the diketide ketoester intermediate produced by the first chain extension is reductively processed prior to the addition of the

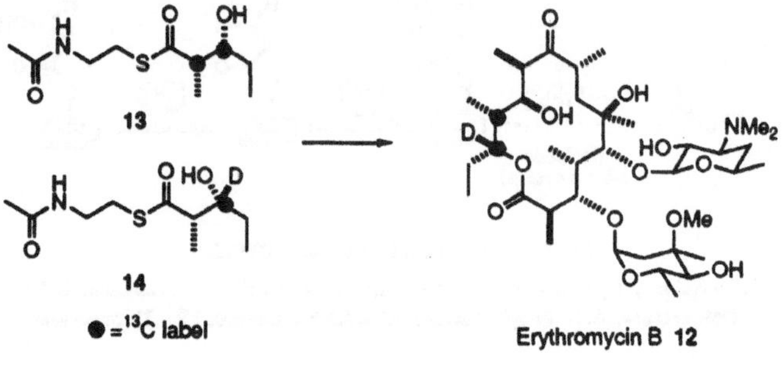

● = ^{13}C label   Erythromycin B **12**

Scheme 4

second chain extension unit. There is also compelling indirect evidence from the genetics of the polyketide synthase (PKS) that each successive new keto group is fully processed to its final reduced state prior to the subsequent chain extension reaction. This "processive mechanism" seems to apply widely for the chain assembly steps of aliphatic polyketide biosynthesis as clearly shown by many similar feeding experiments for other polyketide systems, e. g. the incorporation of 13 and 14 into methymycin in *Streptomyces venezuelae* which demonstrated the similarity of the initial stages of methymycin and erythromycin biosynthesis [25].

2.5
The Reactions of Fatty Acid Biosynthesis

All early attempts to isolate active PKS proteins from organisms producing aliphatic polyketides proved unsuccessful. Speculations concerning their mode of action were therefore guided by the strong parallel with the fatty acid bio-synthetic pathway which has long been investigated at the enzyme level [26, 27]. Formation of a fatty acid is initiated by condensation of a starter unit (commonly acetate) with an extender unit (malonate) (Scheme 5). The resulting

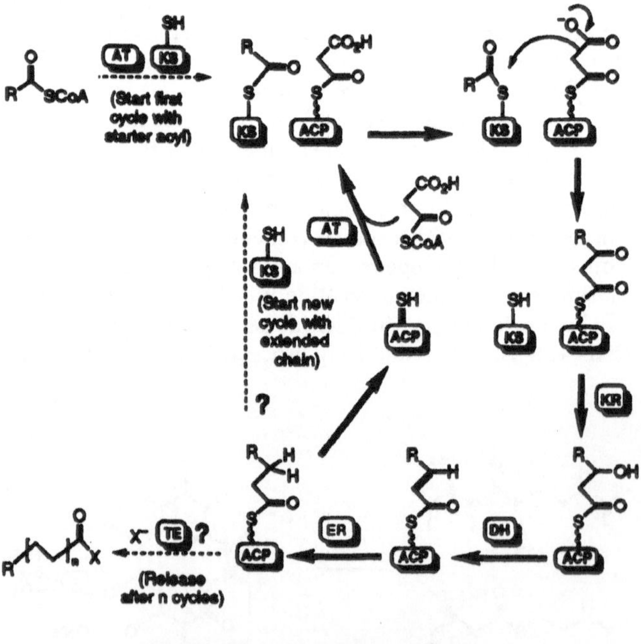

FATTY ACID CHAIN EXTENSION CYCLE

KS = Ketoacyl Synthase; ACP = Acyl Carrier Protein; KR = Ketoreductase; DH =
Dehydratase; ER = Enoylreductase; AT = Acyltransferase; TE = Thioesterase

Scheme 5

β-ketoester is then fully processed (reduced, dehydrated and reduced again) to give an elongated saturated fatty chain. A second chain extension cycle then takes place with the condensation of a new extender unit, followed by complete reductive processing of the new keto group. To carry out this repetitive sequence of operations the fatty acid synthase (FAS) requires a minimum set of catalytic activities, with each activity being responsible for one step of the cycle of reactions, i.e. acyltransferase (AT), β-ketoacyl synthase (KS) and acyl carrier protein (ACP) for chain elongation; β-ketoreductase (KR), dehydratase (DH), and enoyl reductase (ER) for processing of the β-keto group; and thioesterase (TE) for hydrolytic release of the full length chain. Remarkably, a typical FAS requires only a single set of these enzymes to carry out all the successive chain extension cycles. Control over chain length is achieved by finely tuned substrate specificity of key enzymes, notably the KS and TE components.

The constituent proteins of a FAS can be organised in markedly different ways depending on the organism. In bacteria, the individual activities (domains) are freely dissociable and can be isolated separately [28]. In animals and yeast the domains are bound together by peptide links to form large highly organised non-dissociable multi-domain proteins [26, 27]. The multidomain proteins are designated "Type I" and the dissociable synthases are designated "Type II". The same nomenclature has been adopted in the PKS field.

2.6
Sequencing of the Genes Coding for the Erythromycin PKS

The first clue to the structure of the erythromycin PKS came with the sequencing of the corresponding genes. These were located in the vicinity of the gene coding for erythromycin resistance *ermE*. This method was pioneered by Hopwood and Sherman who used it to identify the genes responsible for the actinorhodin PKS [29].

On both sides of *ermE* there are regions of DNA with open reading frames thought, on the basis of sequence homology with known enzymes, to encode for non-PKS proteins responsible for the late stages of the biosynthesis, from 6-deoxyerythronolide B 7 to erythromycin A 5. Many of these assignments have been confirmed by targeted disruption of these genes to give mutants from which late intermediates have been isolated (see above). A map showing these regions of the genome is presented in Fig. 1.

Sequencing further away from the resistance gene, the Leadlay group in Cambridge [30] and Katz and co–workers [31] at the Abbott Laboratories jointly discovered three exceptionally large open reading frames, each containing about 10 kbp of DNA designated *eryAI, AII* and *AIII*. Each of these genes codes for a giant (> 3000 amino acids) multifunctional protein. It was proposed therefore that the erythromycin PKS consists of three giant proteins called DEBS 1, DEBS 2 and DEBS 3 respectively [30 – 34]. In the primary sequence of the putative proteins it was possible to identify regions with motifs showing strong sequence homology with the active site motifs of proteins involved in fatty acid synthases. It was this homology that convinced the investigators that they had discovered the genes coding for the elusive PKS proteins.

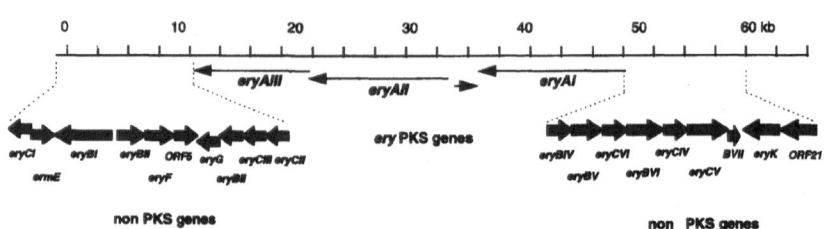

Fig. 1. Map of the *S. erythraea* genome containing genes associated with the late stages of the erythromycin biosynthetic pathway

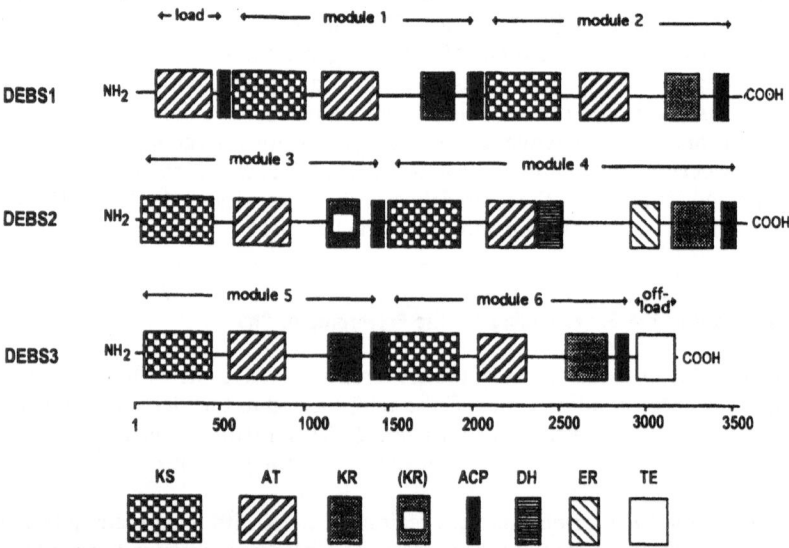

Fig. 2. Predicted domain organisation of the DEBS Proteins. Ketoacyl Synthase (KS); Acyl Transferase (AT); Dehydratase (DH); Enoyl Reductase (ER); Keto Reductase (KR); Acyl Carrier Protein (ACP); Thioesterase (TE). Each domain is represented by a *box* with *coded shading* whose length is proportional to the size of the domain; (KR) indicates an inactive KR domain. The *ruler* indicates the residue number within the primary structure of the constituent proteins. Linker regions are shown in proportion

In the linear representations of the primary sequence of the proteins shown in Fig. 2, sections thought to be associated with specific catalytic activities are indicated by coded blocks. These catalytic centres are termed *domains*. Each domain is thought to be folded to form a localised globular structure with a catalytic role dictated by its specific active site. The various regions between the domains which are unmarked in this diagram are thought to play a passive but vital structural role in maintaining the various domains in the correct topology for co-operation in the overall catalytic process. Most of these structural regions consist of short sections of amino acids rich in alanine, proline and charged residues, up to 30 residues long, which are thought to serve as linkers between the globular domains. There are also intriguing much longer sections of primary sequence in all three proteins preceding the KR domains which cannot be associated with any specific catalytic activity.

2.7
Organisation and Function of the Erythromycin PKS

To help explain the synthetic implications of this organisation of the domains a different representation of the PKS is shown in Scheme 6, in which domains are represented by circles and the linker regions and structural residues are ignored. In all there are six ketosynthase (KS) domains, one for each chain extension cycle. Following each KS there are two domains, AT and ACP, essential for the condensation step, plus an appropriate set of optional domains KR, DH, ER, for modification of the new keto group (except module 3). To aid the analysis, the optional domains are raised above the line of the essential domains in this cartoon version of the structure. It can be seen that the activities are organised into six "*modules*" (two per protein), each of which is able to carry out a condensation step using the three essential domains followed by modification of the keto group in the newly formed keto ester to some extent (except module 3). There is a striking correspondence between the set of domains in successive modules and the extent of keto group modification in the newly added C_3 unit of successive intermediates. At the start of the first protein, DEBS 1, there is a "loading" module for the starter acid (propionate) which consists of an AT linked to an ACP. Module 6 is followed by a thioesterase domain which is thought to catalyse lactonisation of the polyketide chain and so release the completed macrolide structure from the enzyme.

According to this modular analysis, each protein catalyses two cycles of chain extension. The term "*cassette*" has been proposed for the giant proteins [34]. All three cassettes in the erythromycin cluster are bimodular, but in other systems, such as the rapamycin [35] and tylosin [36] PKSs, the size of a cassette can vary from one to six chain extension modules. The three cassettes co-operate in some way to form an extraordinarily complex molecular assembly line. The biosynthetic intermediates remain PKS-bound throughout the whole synthetic sequence via *thioester* links. A challenging feature of this organisation is the mechanism which controls the ordering of the cassettes in the assembly line so that the transfer of the growing chain from one cassette to the next is correctly controlled.

The correspondence between the domain composition of successive modules and the structure of each newly added C_3 unit in successive intermediates is persuasive but it does not prove that the modules function in the order indicated. The protein could fold in many different ways to bring sets of non-contiguous domains together to form functional modules. It was a high priority to establish that modules are indeed made up of adjacent sets of domains as indicated, and also that the individual modules and domains are used in the order shown.

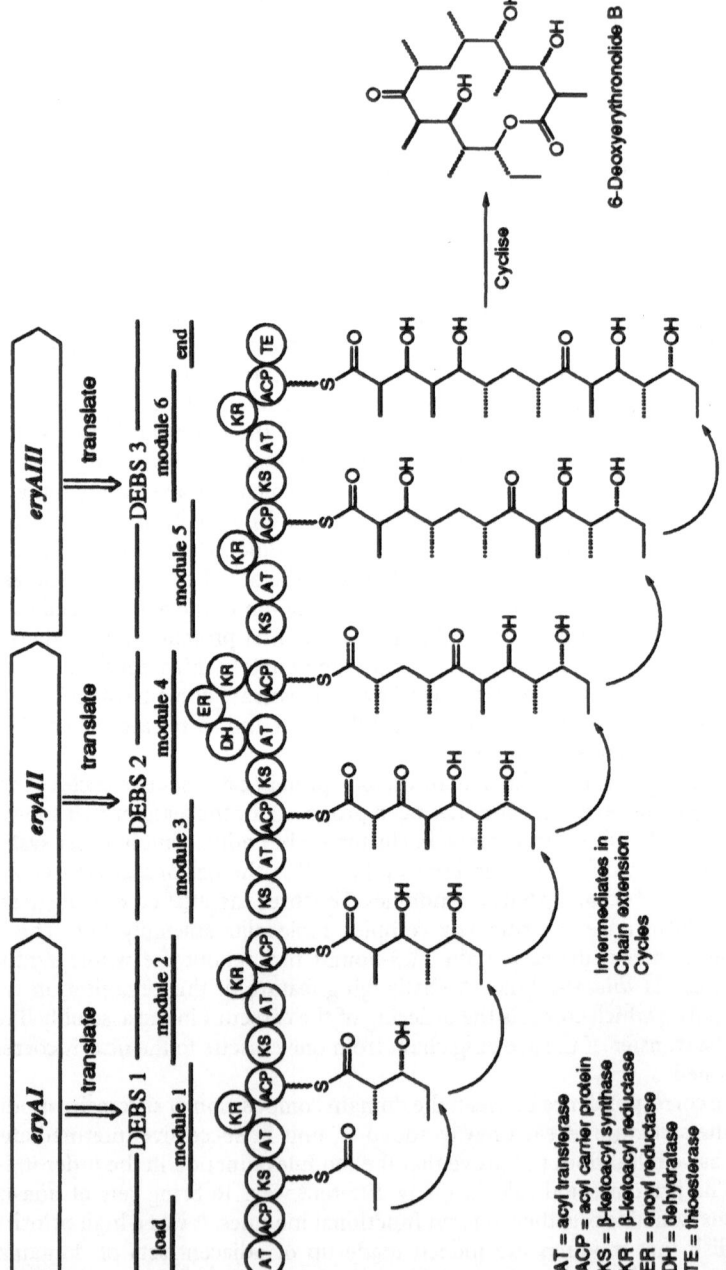

AT = acyl transferase
ACP = acyl carrier protein
KS = β-ketoacyl synthase
KR = β-ketoacyl reductase
ER = enoyl reductase
DH = dehydratase
TE = thioesterase

Scheme 6

The first strong supporting evidence came from Katz and co-workers who carried out pioneering gene disruption experiments. In 1991 they disrupted the β-ketoacyl reductase (KR) in module 5 and from the resulting mutant they isolated a partly processed erythromycin analogue 15, in which a keto group has survived at the predicted position of the macrolide in place of the normal hydroxyl group [37]. The presumed product of the PKS is therefore the 5-keto analogue 16 of 6-deoxyerythronolide B (Scheme 7). More recently the enoyl reductase (ER) in module 4 was disrupted and an analogue 17 of erythromycin was isolated containing a double bond at C-6/C-7 (Scheme 8) [38]. It can be concluded that the disrupted keto reductase operates in the fifth cycle of chain extension and that the disrupted enoyl reductase operates in the fourth cycle as predicted by the simple successional arrangement of modules in Scheme 6.

These two experiments also had important implications for genetic engineering of polyketide biosynthesis to produce novel metabolites. The formation of the two structural analogues of the polyketide chain and the fact that they are correctly lactonised shows that the structure of the nascent chain may not play a critical role in polyketide biosynthesis, and at least some altered polyketides can be substrates for further cycles of chain extension. This implies that the correct transfer of the growing chain from one synthase unit to the next may reside more in the specific juxtaposition of the various domains rather than in the conventional substrate specificity of a particular synthase domain for the structure of the approaching chain. It is interesting and encouraging that the product released from the PKS in both experiments was at least partly processed towards an analogue of erythromycin A, showing that the elabora-

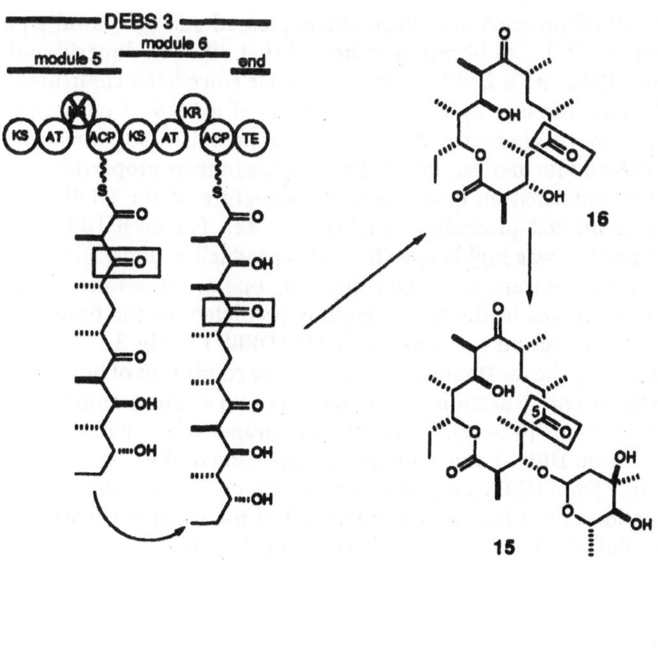

Scheme 7

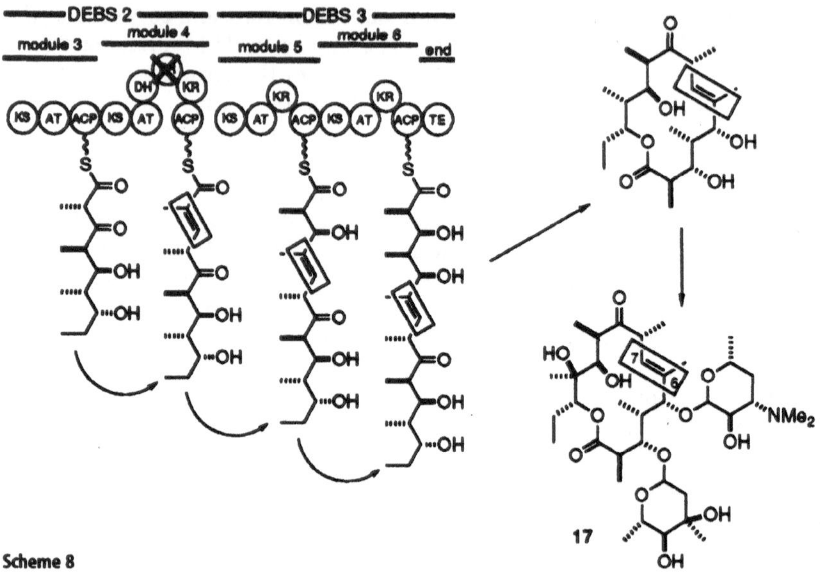

Scheme 8

tion enzymes also possess a useful degree of relaxation in their substrate specificity.

2.8
Isolation and Structural Studies of the Erythromycin PKS

All three DEBS proteins have been overexpressed and successfully purified to homogeneity [39]. Gel filtration indicated that all were dimeric under native conditions [39], and a DEBS multienzyme has since been confirmed to be exclusively homodimeric under the conditions of analytical ultracentrifugation [40]. The dimeric character of the PKS proteins mirrors the behaviour of the animal FAS and further reinforces the parallels in their properties.

The first information concerning the structure of the DEBS homodimers came from limited proteolysis studies [40–42]. For each DEBS protein the cleavage pattern was highly specific and reproducible under given conditions. Sets of fragments were obtained containing one or more intact domains produced by cleavages in the linker regions predicted on the basis of sequence alignments. The results are summarised for DEBS 1 in Fig. 3.

Significantly, the most rapid cleavages under conditions of limited proteolysis were always found to occur between the two chain extension modules housed in each multienzyme [40–42]. Other initial cleavages split the N-terminal loading didomain from DEBS 1 and a didomain consisting of the C-terminal offloading thioesterase from DEBS 3 together with its attached ACP domain [39, 40]. With the exception of the loading didomain, all of the fragments corresponding to intact modules behaved as *homodimers* on gel filtration.

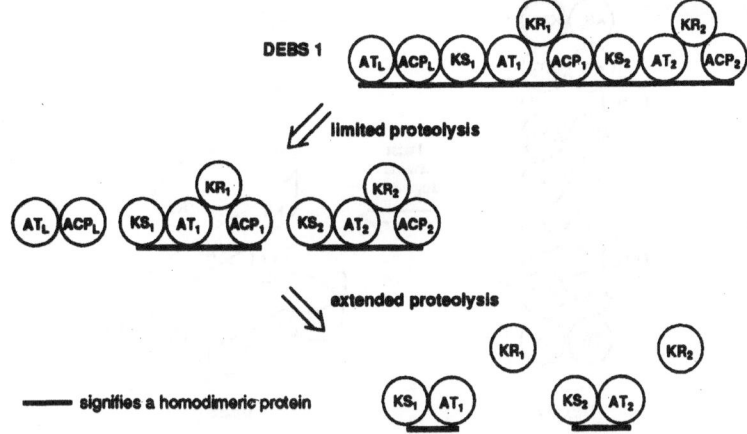

Fig.3. Pattern of fragments generated by controlled proteolysis of DEBS1

More extensive proteolysis of DEBS 1, DEBS 2 and DEBS 3 gave smaller fragments [40–42]. Most significantly, the six KS domains were released as didomains with their attached AT domains and all of these were *homodimeric*. In contrast, fragments consisting of single domains associated with keto group modification were all *monomeric*: the KR domains of modules 1, 3, 5 and 6, and the single ER domain from module 4.

Proteolysis of DEBS dimers in which the thiols at the active sites of co-operating ACP and KS residues had been crosslinked by prior treatment with 1,3-dibromopropanone gave a different pattern of fragments consistent with the formation of crosslinks *between* two complementary protein chains [40]. It was therefore concluded that the KS of one chain co-operated with the ACP residing in the complementary chain rather than in its own, as had been demonstrated earlier for the animal FAS.

The Cambridge group which carried out these experiments rationalised the results in terms of a dimeric "double helical" structure for the DEBS proteins, illustrated for DEBS 3 in Fig. 4 [40]. To arrive at the double helix the two poly-peptide chains of the DEBS are first folded to give domains and then lined up side by side in a parallel "head to head", "tail to tail" fashion as shown. The two chains are then twisted together to form a double helix with the KS, ACP and AT domains of each module forming the core of the structure. The helix is shown in ribbon form to aid clarity in the positioning of the domains. The sense of the helical twist is arbitrary at this stage. The twist brings the ACP of one chain below the KS domain of the identical module in the complementary chain allowing the observed inter-chain cross-linking.

The optional reductive domains in each module needed for ketone group modification (DH, ER and KR) form loops that protrude out sideways from the central core, while maintaining the active sites of all reductive domains within range of the phosphopantetheine group on the adjacent ACP in the *same* chain. There is therefore no contact between reductive domains of complementary

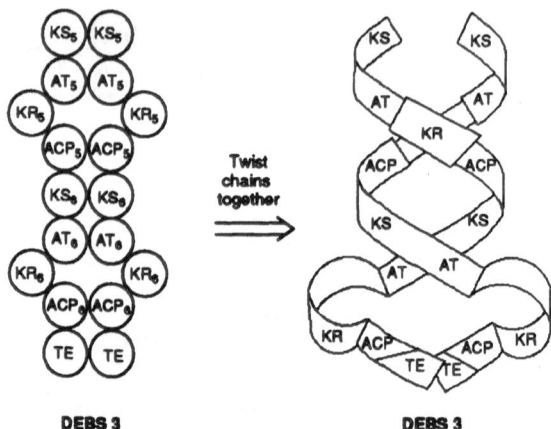

Fig. 4. Proposed folding of DEBS 3 to give the "double helical" model

chains which is consistent with the monomeric nature of these domains in the isolated state (see Fig. 3). In contrast the KS domains (and TE domain of module 6) are in intimate contact in the core of the helical structure and so it is reasonable that the proteolytic fragments containing them would retain the homodimeric character of the intact protein. Three such double helices stacked on top of the other give rise to the complete erythromycin PKS.

Further evidence concerning the arrangement of domains of the DEBS homodimers came from a complementation study in which two genetically engineered versions of the protein were made, one defective in the KS domain of module 1, the other defective in the KS domain of module 2 [43]. Neither protein was active but activity could be restored by preparing a heterodimer containing one chain of each type. The work also established that a given ACP co-operates with KS domains in complementary chains. An alternative model for the erythromycin PKS has been proposed in which the homodimer is formed with the complementary chains running *anti*parallel side by side. The resulting flat modules are then stacked one above the next as in a stack of pancakes with linker regions running vertically between linked modules at the *outside* of the stack (Fig. 5).

The architecture of the homodimers is crucial to understanding how these fascinating synthases operate. This subject is bound to be a high priority for future work both in the PKS and the animal FAS fields.

2.9
Genetic Engineering of the Erythromycin PKS

The pioneering studies by the Katz group have already been discussed in support of the proposed modular structure of the PKS. A common characteristic of these genetic engineering experiments was the essentially destructive character of the mutations. The activity of individual domains could be destroyed without

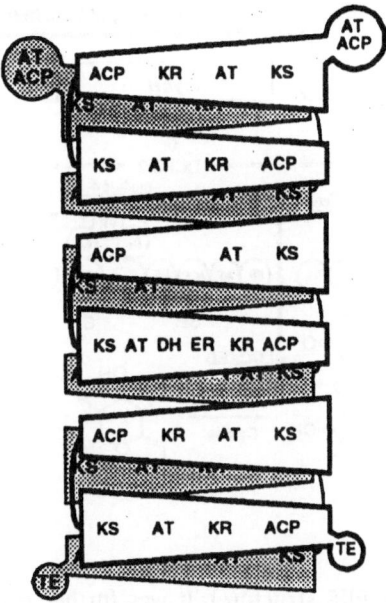

Fig. 5. Proposed model for DEBS based on the proposed linear "head to tail" model for the animal FAS

destroying the overall capacity of the PKS to make macrolide structures. This strategy could therefore be used to generate novel products, but clearly the range of possible variations was very limited.

The erythromycin PKS has more recently been genetically engineered in many different ways. Most work to date has relied upon an expression plasmid pRM5 developed by Hopwood and co-workers for expression of genes in *Streptomyces coelicolor*, an organism which normally produces an aromatic polyketide, actinorhodin [44]. In initial experiments it was demonstrated that the whole of the erythromycin PKS could be transferred to *S. coelicolor* to give a mutant with the capacity to produce the unelaborated macrolide product, 6-deoxyerythronolide B 7 as well as the analogue with an acetate starter unit [45]. This mutant has also proven to be a more efficient vehicle for incorporation experiments with added precursors than the natural host, *S. erythraea* [46]. The development of this highly efficient vehicle for genetic engineering led to rapid progress in the production of modified polyketide products.

The Leadlay Staunton group made the first significant breakthrough by repositioning a domain within the PKS to show that the repositioned domain could carry out its normal type of reaction in its new context on a foreign substrate. This entailed adding a copy of the thioesterase (TE) domain to the terminus of DEBS 1 [47]. The aim was first to prevent further chain extension by blocking the docking of the C-terminus of DEBS 1 onto the N-terminal end of DEBS 2 (this was the predicted consequence of this change according to the

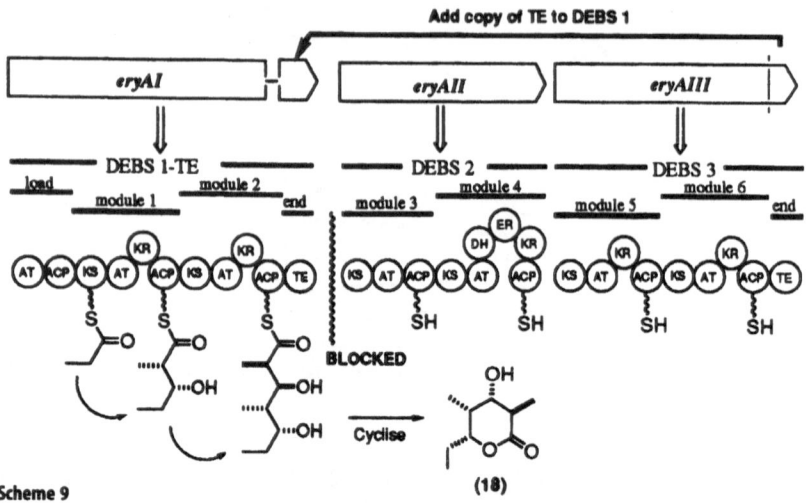

Scheme 9   **(18)**

'helical' model for the PKS structure). It was further reasoned that the TE domain would play an *active* role in the engineered protein by catalysing the release of the triketide intermediate from module 2 as a δ-lactone.

A mutant of *S. erythraea* containing the engineered protein (called DEBS 1-TE) (Scheme 9) did give the anticipated δ-lactone 18. As expected, erythromycin production was shut down completely. A vital control experiment was then carried out to establish the second point – that the relocated domain was acting as an *active* agent of chain release rather than just a passive block to further chain extension. A further mutant was made in which a genetically engineered *dead* copy of the TE was placed at the end of DEBS 1. The inactivated TE differed from the normal enzyme only in the replacement of the key serine residue at the active site by an alanine; this conservative change should not alter the folding of the protein significantly. This control mutant did not produce erythromycin showing that the block to further transfer of the chain still existed. The δ-lactone 18 was produced but in much reduced yield, demonstrating that the TE in the first mutant was playing an active role in chain release.

Further examples of this strategy have since been reported by Cane et al. Their first experiment mirrored the design of DEBS 1-TE in that it involved placing the thioesterase at the end of DEBS 1. The construct was engineered in a different way, however, to yield a protein with a significantly different primary structure, so it is appropriate that it was given a slightly different name: DEBS 1 + TE [48]. The TE has since been relocated to the terminus of modules 5 [48], and 3 [49]. Products consistent with truncation of chain extension at the expected stages were obtained from both mutants (Scheme 10). The truncation at module 5 was especially significant because it led to release of the hexaketide intermediate as a 12-membered ring macrolide 19. The truncation after module 3 caused the formation of two tetraketide products, 21 and 22; the second is

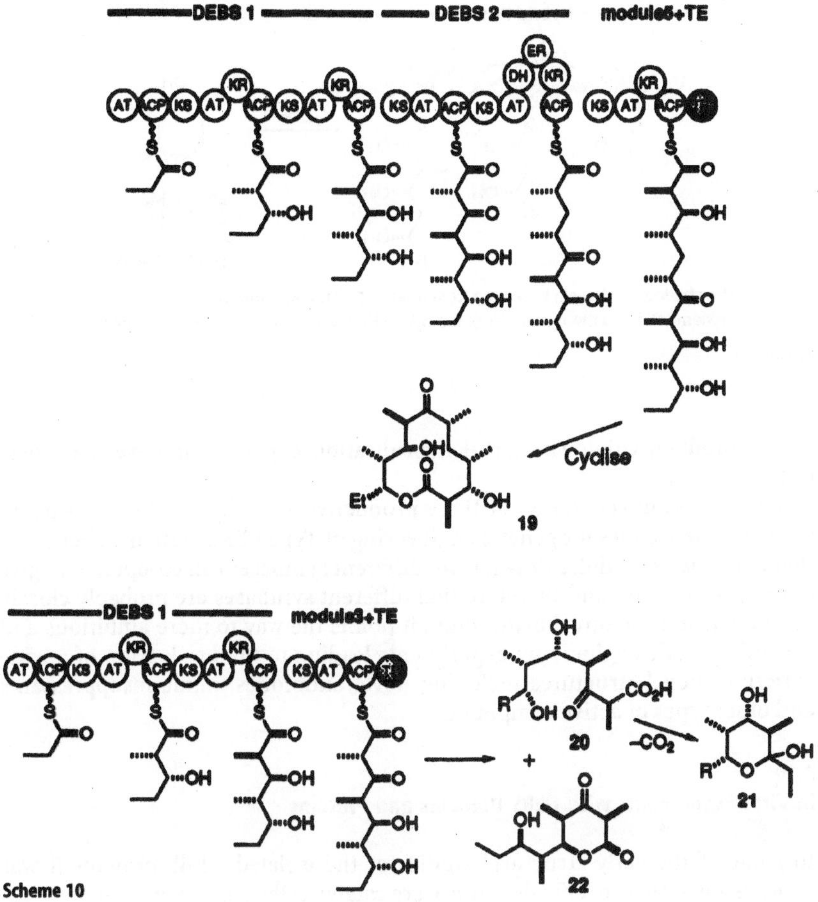

Scheme 10

possibly formed from a ketoacid **20** by decarboxylation after release from the enzyme.

A potentially more versatile strategy for generating novel products is the transfer of domains between foreign PKS clusters. This would amount in effect to "spare part surgery" or "mix and match" swapping of structural residues between different natural products. In the first demonstration of the feasibility of this radical concept, sections of two different *Type I* PKS systems (rapamycin and erythromycin) were hybridised to produce a *hybrid Type I* PKS (Scheme 11) [50]. The experimental model was DEBS1-TE and the experiment involved replacing the AT of module 1 by an AT derived from the rapamycin PKS (this will be described later). The transplanted AT specifies a malonate unit as chain extender rather than methylmalonate. δ-Lactones **23** and **24** were produced lacking a methyl group at C-4, but otherwise the normal structure was maintained at C-2, C-3 and C-5. The *hybrid* PKS therefore produced the predicted

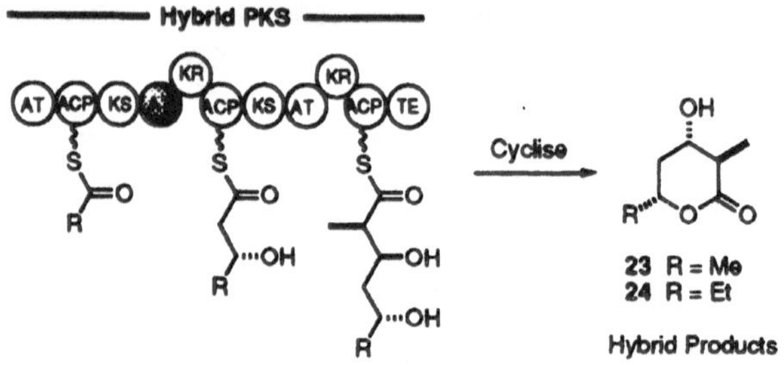

The shaded AT domain is derived from module 2 of the rapamycin PKS
(Scheme 27) and the remainder of the hybrid PKS is as in DEBS 1-TE (Scheme 9)

Scheme 11

hybrid product. Other successful hybridisation experiments have been described [36, 51].

The successful generation of these productive, truly hybrid, PKSs opens up exciting new avenues for genetic engineering of Type I PKS systems. They show that domains or modules drawn from different synthases can co-operate to give a working synthase and therefore that different synthases are probably closely compatible in structure. Furthermore it points the way to more ambitious and versatile genetic engineering experiments leading to the production of a wide variety of novel structures including novel antibiotics, immunosuppressants and other types of active compound.

2.10
In Vitro Experiments with DEBS Domains and Proteins

In some of the early structural studies of the isolated DEBS proteins it was demonstrated that specific domains were catalytically active. For example, cleavage of DEBS 1 with trypsin gave the N–terminal didomain together with two other fragments, comprising modules 1 and 2 respectively. It was found that the N-terminal didomain was specifically radio-labelled with [14C]propionyl–CoA, after incubation, providing the first evidence for its proposed role as the "loading module" for the propionate starter unit. In contrast, modules 1 and 2 were specifically acylated with [14C]methylmalonyl–CoA, indicating that the other two acyltransferases were enzymatically active after proteolysis [41, 42]. Using intact DEBS proteins it was discovered that all the AT domains of the chain extension modules specifically catalysed the hydrolysis of (S)-methylmalonyl-CoA but were inert to the (R)-isomer. This was the first evidence that the (S)-isomer of methylmalonyl-CoA serves as building block for both chain extension modules of DEBS 1 (see later) [42].

In 1995 the engineered, self-sufficient PKS DEBS 1-TE was isolated in pure form [52]. When incubated with the appropriate building blocks, propionyl-CoA and methylmalonyl-CoA, and a reducing agent, *specifically* NADPH, the protein catalysed the formation of the δ-lactone triketide product **18** (Scheme 12). This

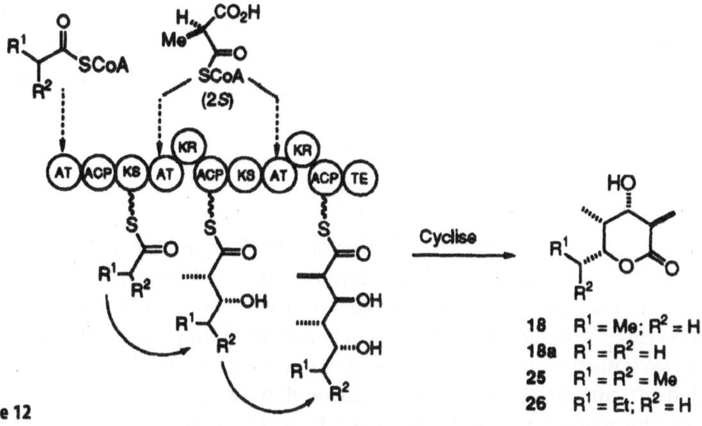

Scheme 12

18	$R^1 = Me$; $R^2 = H$
18a	$R^1 = R^2 = H$
25	$R^1 = R^2 = Me$
26	$R^1 = Et$; $R^2 = H$

was the first demonstration of product formation in vitro by any modular PKS. Given the high state of purity it was established that the PKS could carry out its normal function without assistance of external proteins. The PKS showed a more relaxed degree of substrate specificity in its loading domain than was apparent in vivo. n-Butyryl- and isobutyryl-CoA esters were acceptable as starter acyl groups in vitro, to give novel products 25 and 26 as well as the acetyl and propionyl starter acyl groups found to be effective in vivo (see Scheme 12).

A mechanistically important result to emerge from this study was the rigorous determination of the stereochemistry of the methylmalonate units used for chain extension in modules 1 and 2. This was of interest because the two branching methyl groups in the lactone product have opposite stereochemistries. In principle it was possible that either the (R)-isomer of the methylmalonate was incorporated at one site and the (S)-isomer at the other by a common incorporation mechanism, or that a single enantiomer was incorporated at both sites in which case the two modules would have to follow different sequences of reactions. It was unambiguously shown that the (S)-isomer served for both chain extensions. It is unlikely, therefore that altered products with *different stereochemistry* will be generated by transferring AT domains and some other means will have to be devised to achieve this desirable end.

Similar results have come from studies of the DEBS 1+TE system in vitro [53]. When DEBS 1 + TE was incubated with appropriate precursors in the absence of the reducing agent NADPH the keto 27 and the pyrone 28 analogues of the triketide lactone were obtained (Scheme 13) [53, 54]. This latter result resembles the similar formation of a pyrone product by the animal FAS and the 6-methylsalicylic acid PKS under similar conditions [55]. Clearly, the β-ketoester intermediates can be transferred from one module to the next without further processing, even in modules which are geared to the production of a hydroxyester intermediate.

Other significant in vitro studies include the demonstration that a cell free extract of all three DEBS proteins expressed heterologously in *S. coelicolor* was

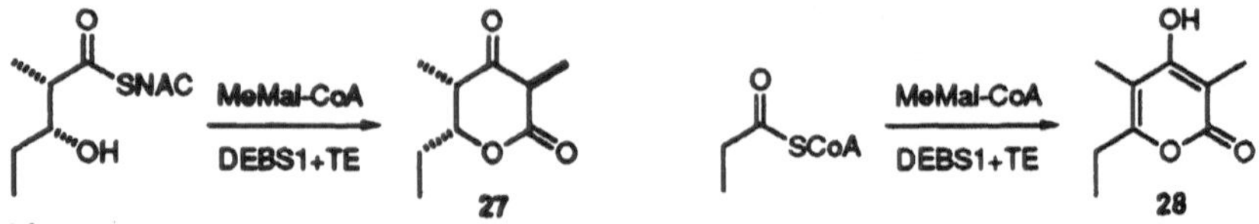

Scheme 13

capable of synthesising 6-deoxyerythronolide B 7, albeit with rather low efficiency [56]. The engineered protein module3+TE has been purified to homogeneity [57]. In collaboration with a pure extract of DEBS 1 this produced the tetraketide products 21 and 22 (see Scheme 10). It may be possible to use one of these systems to catch the two cassettes in contact as the developing chain is passed from one to the other by treatment with appropriate cross-linking reagents. Unfortunately, attempts so far have not met with success and the nature of cassette docking in multi-cassette PKS systems remains a mystery.

The substrate specificity of the thioesterase domain is of crucial concern in the quest for a rational basis for genetic engineering of PKSs to produce novel products: it would be futile to engineer a PKS system to produce a particular novel product if there is no effective mechanism for its release from the final ACP. The didomain from the end of DEBS 3 was over-expressed in *E. coli* to give milligram amounts of protein [58]. The resulting protein was shown by electrospray mass spectrometry to have the correct molecular weight for the didomain with an *apo*-ACP in which the phosphopantetheine group had not been added. This deficiency did not matter as far as the planned studies of the TE specificity were concerned. In the preliminary work it was shown that the protein was able to bind *p*-methylphenylsulphonyl fluoride (PMSF), a standard inhibitor of chymotrypsin-like enzymes, to the hydroxyl of a specific serine residue located in the putative active site of the enzyme. This evidence strongly supported a proposed cleavage mechanism involving initial transfer of the developed acyl chain from the thiol of the neighbouring ACP to the hydroxyl group of an active serine residue on the TE (Scheme 14). The resulting oxygen ester is then cleaved by attack of a suitable nucleophile to release the product. In forming a macrolide ring (the presumed normal role of the thioesterase), the nucleophile would be the appropriate hydroxyl group at the distant terminus of the polyketide chain.

To assess its substrate specificity, the thioesterase was next challenged with synthetic acylesters derivatised as *p*-nitrophenyl and *N*-acetylcysteamine (NAC) esters (exemplified in Scheme 15) [59]. A wide range of structurally varied substrates with both leaving groups was cleaved. By electrospray mass spectrometry it was shown that acyl enzyme intermediates were formed as predicted. The resulting acyl enzyme intermediates could be cleaved either by water to give the corresponding acid, or by an added alcohol such as ethanol to give an ester. None of the substrates tested gave any macrocyclic lactone even when there was a hydroxyl group elsewhere in the chain at a suitable distance from the acyl terminus. The work established that the TE has alternative mechanisms for product release (hydrolysis or ester formation) when lactonisation is not pos-

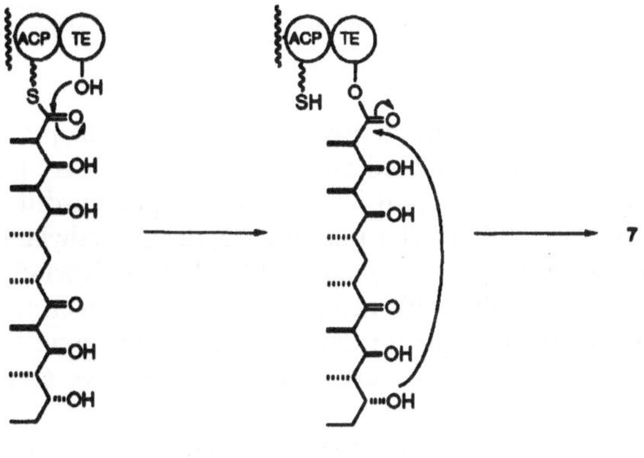

Scheme 14

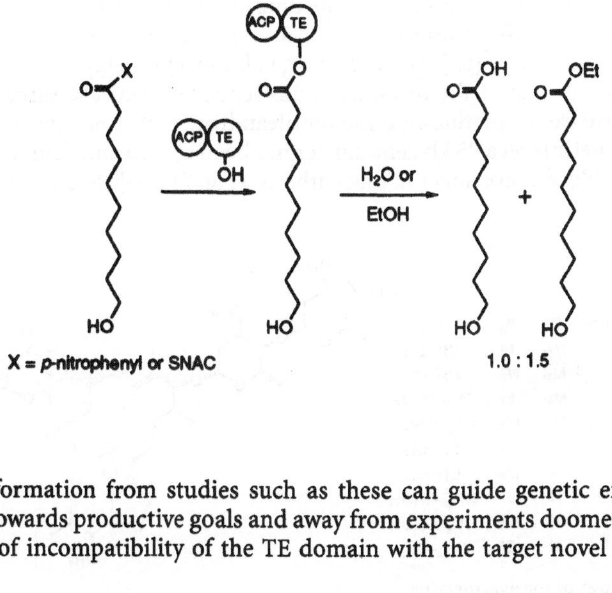

X = p-nitrophenyl or SNAC   1.0 : 1.5

Scheme 15

sible. Information from studies such as these can guide genetic engineering studies towards productive goals and away from experiments doomed to failure because of incompatibility of the TE domain with the target novel polyketide product.

2.11
Summary

The erythromycin PKS provides the structural and mechanistic paradigm against which other synthases will be compared both in vitro and in vivo. Equally important are the impressive genetic engineering technologies which have been developed for modification of this synthase and which should be more widely applicable to other synthases. Despite these rapid advances, there remain enormous gaps in our understanding of the structure and function of these challenging systems.

3
The Avermectins

The naturally occuring avermectins are a family of eight, pentacyclic structures containing a sixteen-membered macrolide ring which are produced by *Streptomyces avermitilis* (Fig. 6) [60]. These eight compounds differ in structure due to the variability of PKS starter unit at C-25, the O-methylation pattern at C-5, and the hydration-dehydration pattern at C-22/C-23. Other closely related natural products include the milbemycins and the nemadectins. The avermectins exhibit potent broad spectrum antiparasitic activity, and the semi-synthetic Ivermectin **29** (hydrogenated avermectin B1) displays even greater efficacy [60, 61].

Feeding studies have shown the avermectin aglycone to be derived from seven acetate and five propionate units [62]. The isobutyrate and 2-methylbutyrate starter units derive from catabolism of L-valine and L-isoleucine respectively [63]. [^{18}O]Acetate and propionate feedings have shown that all the oxygen atoms other than those attached to C-6 and C-25, are derived from their carboxylate precursors [62]. The C-25 oxygen is most probably derived from the starter unit and that of the furan from molecular oxygen. The three methoxy groups derive from L-methionine and the oleandrose units from glucose [63, 64].

At the genetic level a 95 kb gene cluster has been cloned and demonstrated to be responsible for avermectin biosynthesis (Fig. 7) [65]. Sequencing of the

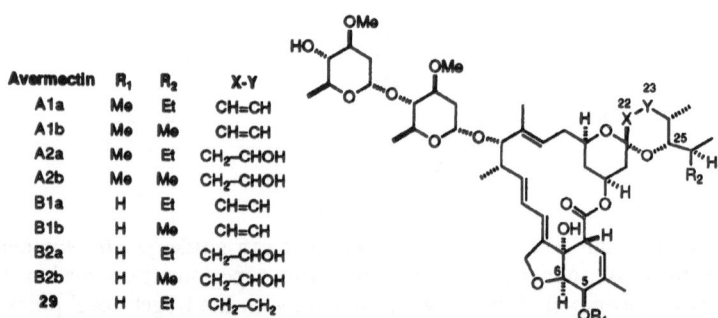

Avermectin	R₁	R₂	X-Y
A1a	Me	Et	CH=CH
A1b	Me	Me	CH=CH
A2a	Me	Et	CH₂-CHOH
A2b	Me	Me	CH₂-CHOH
B1a	H	Et	CH=CH
B1b	H	Me	CH=CH
B2a	H	Et	CH₂-CHOH
B2b	H	Me	CH₂-CHOH
29	H	Et	CH₂-CH₂

Fig. 6. Structures of the avermectins

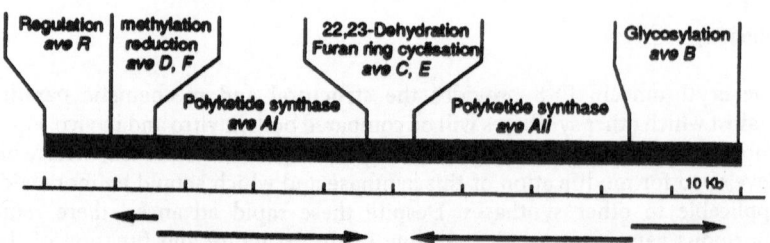

Fig. 7. Organisation of the genes for avermectin biosynthesis

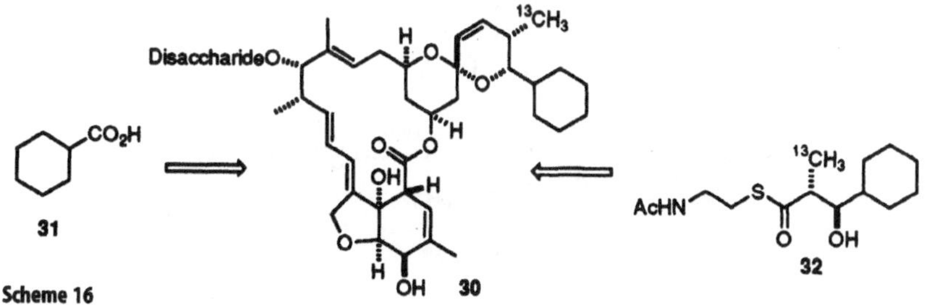

Scheme 16

cluster revealed structural organisation for a type I PKS with strong similarity to that of the erythromycin PKS. A central 70 kb fragment contains two convergent transcripts, *aveAI* and *AII*, which appear to encode multifunctional proteins with six modules present in each transcript. The first module of *aveAI* demonstrates the presence of a loading didomain and first extension module as observed for the erythromycin PKS. Further genes responsible for post PKS modifications and regulation both flank and occur between the two PKS genes.

The range of naturally occurring avermectins has been extended by utilising a "mutasynthesis" approach [66]. The *S. avermitilis* blocked mutant employed lacked a branched chain decarboxylase activity which is responsible for production of the PKS starter acids. Feeding of the "natural" starter acids restored the production of natural avermectins, whereas some 40 "unnatural" starter acids led to many novel avermectin analogues modified at C-25. Doramectin **30**, an extremely effective antiparasitic treatment marketed as Dectomax, was isolated after feeding cyclohexane carboxylic acid **31** (Scheme 16). The NAC thioester analogue **32** of the diketide biosynthetic intermediate was also incorporated into **30** using this mutant strain (see Scheme 16) [67]. This example, and those for the tetronasin producer discussed later, are the only examples to date of the incorporation of an analogue of a natural PKS-bound precursor.

The avermectin PKS loading module has been used to generate a hybrid PKS system by replacing the loading module of DEBS 1-TE. The new hybrid PKS produced new hybrid polyketide (triketide lactones) which incorporated the isobutyrate and 2-methylbutyrate starter acids of avermectin biosynthesis, as well as the normal acetate and propionate starter units of erythromycin biosynthesis [51].

4
Tylosin

Tylosin **33** from *Streptomyces fradiae* is a representative member of the large family of 16-membered macrolide antibiotics commonly utilised in veterinary medicine. The aglycone core, tylactone **34** is constructed from two acetate, five propionate and one butyrate units [68, 69]. The processive mechanism for its biosynthesis was demonstrated by the pioneering contributions of Hutchinson

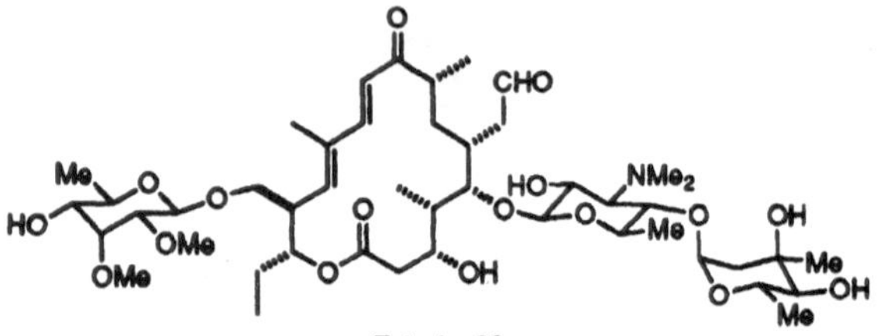

Tylosin 33

and co-workers who demonstrated the incorporation of NAC thioesters of di- and triketide intermediates (Scheme 17) [24]. Further supporting evidence comes from the isolation of partially elaborated intermediates of chain elongation in mutants of *S. fradiae* [70].

The gene cluster for tylosin biosynthesis is known [71], and PKS encoding regions for the biosynthesis of both tylactone and the structurally very similar compound platenolide 35 in *Streptomyces ambotaciens* have been discussed [36]. The gene organisation, but not the sequence was published. The platenolide PKS was reported to consist of five ORFs which encode two bimodular and three monomodular multifunctional proteins. The reported domain sequence within the modules matches that required for the platenolide aglycone synthesis, and the first protein contains a loading module consisting of a KS-AT-ACP tridomain. This information was utilised in the construction of a hybrid PKS consisting of the tylactone PKS loading module in place of that of the platenolide PKS. The mutant strain carrying this new hybrid PKS was shown to produce the new hybrid polyketide 16-methyl platenolide 36 as expected.

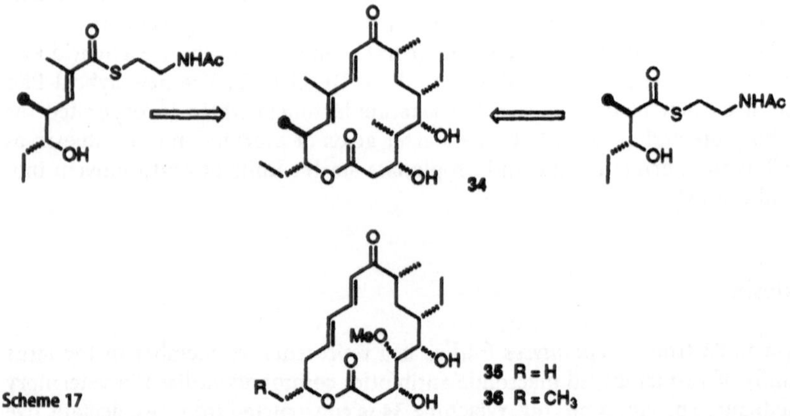

Scheme 17

5
Nargenicin A and the Solanapyrones A and B

Certain cyclic moieties present within some polyketide structures are considered to occur via an intramolecular Diels-Alder reaction. Such a process has been proposed for several metabolites, of which nargenicin A and the solanapyrones A and D are the most thoroughly investigated.

Feeding studies to *Nocardia argentinesis* with ¹³C labelled NAC thioesters of di-, tri-, tetra- and pentaketide intermediates demonstrated regiospecific incorporations into nargenicin 37 (Scheme 18) which are consistent with an intramolecular Diels-Alder cyclisation of a linear nonaketide intermediate (Scheme 19) [72, 73].

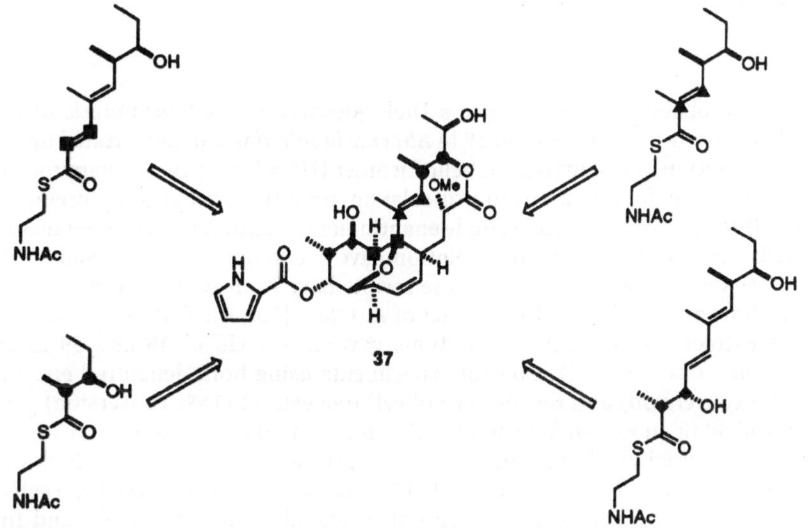

Scheme 18

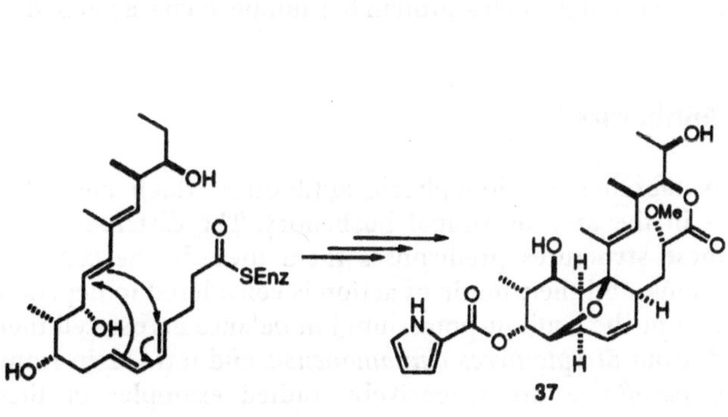

Scheme 19

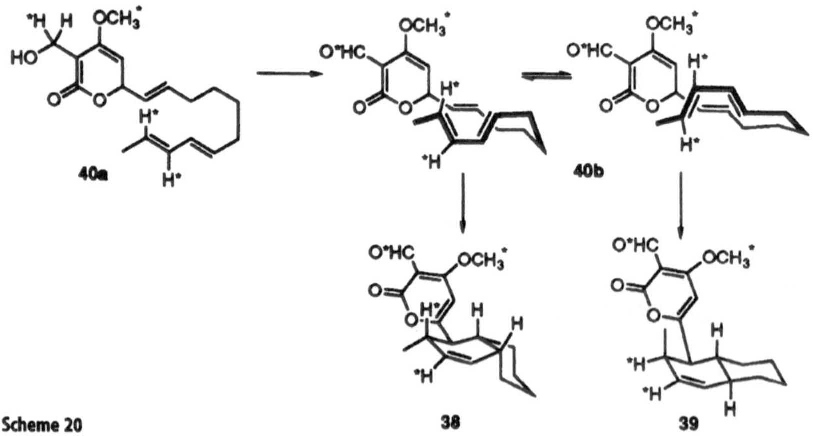

Scheme 20   38   39

The probable intermediacy of a Diels-Alderase in the biosynthesis of the solanapyrones A and D **38** and **39** in *Alternaria solani* was demonstrated by the incorporation of a deuterium labelled isomer (H* = D) of pro-solanapyrone II **40a** (which is first oxidised to pro-solanapyrone III **40b**), yielding predominantly the *exo* isomer **38** as a single enantiomer (Scheme 20) [74]. Spontaneous cyclisation of **40b** in aqueous solution gave predominantly *endo* isomer **39**, which was of course racemic. These results have been reinforced by a study which investigated a cell free extract of *A. solani* [75]. Incubation of **40b** with this extract under specific conditions gave 25% yield of **38** and **39** in an *exo/endo* ratio of 53:47. Control experiments using both denatured enzyme (10% conversion) and the absence of cell free extract (15% conversion) gave **38** and **39** in an *exo/endo* ratio of 3:97. This allowed calculation of a 15% enzyme catalysed yield of **38** and **39** with an *exo/endo* ratio of 87:13, which was further shown to be > 92% ee for **38**. That the biologically observed *exo* selectivity cannot be achieved by chemical means alone is remarkable, and the combined in vivo and in vitro results leave little doubt as to the presence of a Diels-Alderase, although such a protein has not been characterised.

6
Polyether Antibiotics

Polyether metabolites are ionophoric antibiotics which have their major applications in the area of animal husbandry. The distribution of oxygen atoms in these structures predisposes them towards the complexation of alkali metal ions, and their mode of action is considered to depend upon the ability to disrupt the sodium-potassium ion balance across cell membranes. Monensin A from *Streptomyces cinnamonensis* and tetronasin from *Streptomyces longisporoflavus* are extensively studied examples of these metabolites.

6.1
Monensin A

Monensin A **41** is derived from five acetate, seven propionate and one butyrate units, with an extra O-methyl group derived from L-methionine [76, 80]. The NAC thioester of a postulated triketide chain elongation intermediate has been incorporated intact into monensin A by Patzelt and Robinson. This experiment required the extremely careful use of several β-oxidation inhibitors and other additives to minimise precursor degradation [77].

The most intriguing question which remains concerning the biosynthesis of polyether antibiotics such as monensin A is that of the mechanism by which the series of ether rings are formed. In 1965 Westley et al. proposed a mechanism for the biosynthesis of lasalocid and isolasalocid acids from a common diepoxide precursor which differed only in the regiochemistry of the cyclisation process [78]. In 1983 Cane and co-workers extended this idea to monensin A biosynthesis in which an acyclic (post PKS) triene precursor **42** would be converted into triepoxide **43**, mediated presumably by a cytochrome P-450 system, and then a series, or cascade, of intramolecular epoxide ring openings forms, after a final oxidation step, monensin A **41** (Scheme 21) [79]. Indirect evidence to support this hypothesis was demonstrated by [^{18}O]carboxylate and [^{18}O]-molecular oxygen feeding experiments which showed that the oxygen atoms at the postulated epoxide-derived positions (^{18}O) are derived from molecular oxygen [80].

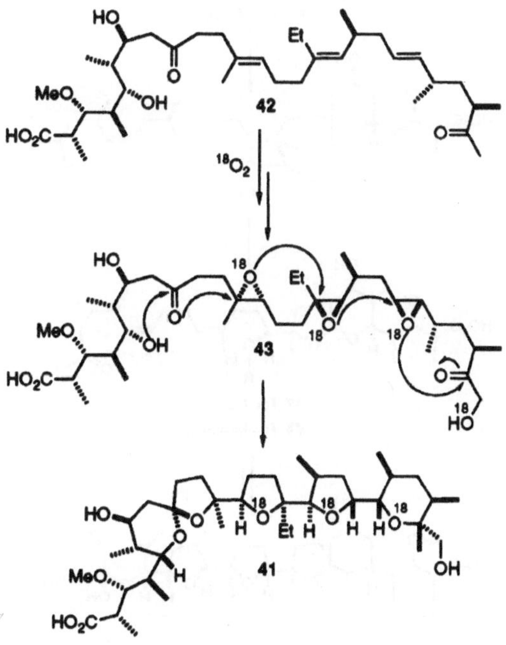

Scheme 21

Biomimetic studies of this proposed cascade have provided further tantalising, albeit incomplete, supporting evidence [81]. Feeding studies with a NAC thio-ester of the required (E,E,E)-triene intermediate 42 in labeled form have so far been unsuccessful, although this may well be due to problems of transport across the cell membrane and degradation by β-oxidation pathways [82].

More recently an alternative biosynthetic pathway has been proposed by Townsend and Basak (Scheme 22) [83]. This alternative model involves the iterative syn-oxidative cyclization of a (Z,Z,Z)-triene intermediate 44 in contrast to the (E,E,E)-triene intermediate 42 of the epoxide cascade mechanism. The putative, alkoxy-linked, non-heme metal-oxo species 45 undergoes a [2+2] cycloaddition to yield the metalloxetane intermediate 46. Reductive elimination then results in an overall syn-addition of the two oxygen atoms 47. Oxidation of the metal intermediate 47 to 48 and two further [2+2] cycloaddition-reductive elimination-metal oxidation cycles would result, after final hydroxylation, in monensin A 41.

In model studies McDonald and Towne have shown that oxochromium reagents can oxidise model (Z)- and (E)-hydroxydienes 49 and 50 to yield

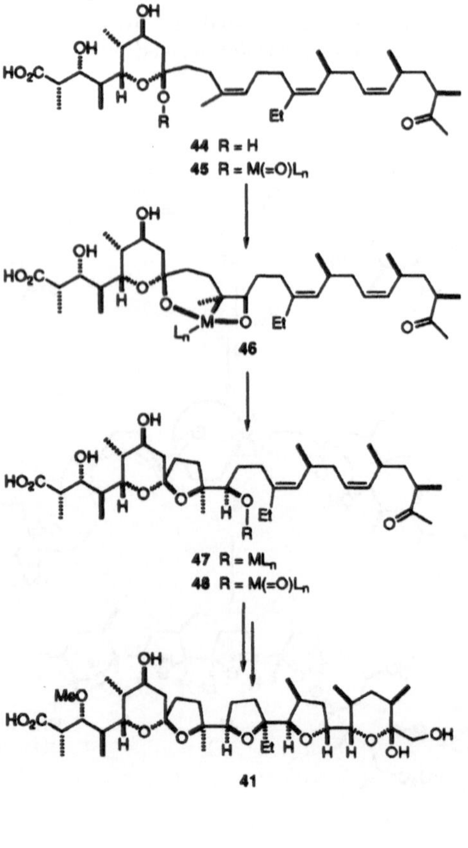

44 R = H
45 R = M(=O)L$_n$

46

47 R = ML$_n$
48 R = M(=O)L$_n$

Scheme 22   41

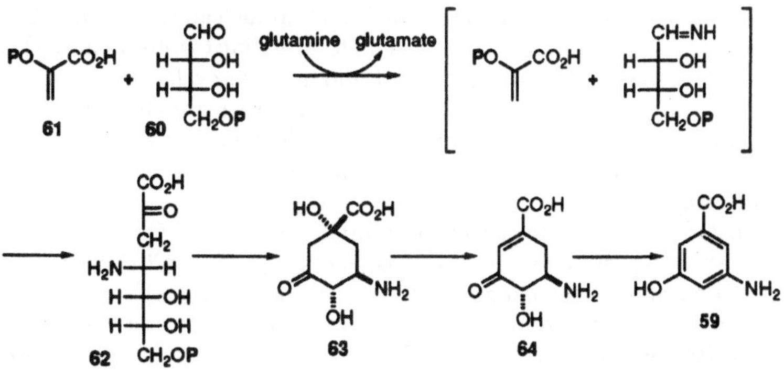

Scheme 25

Evidence for the initial steps of rifamycin B biosynthesis can be seen in the accumulation of a "tetraketide" chain elongation intermediate P8/1-O **58** from a mutant of *A. mediterranei* [92]. This evidence and the demonstration from feeding experiments that the carbons on either side of the ether link in the ansa bridge derive from the same propionate unit [93] indicate that this ether moiety is formed after the initial biosynthesis of a fully extended polyketide chain.

The specific and proximate precursor of the mC_7N unit in ansamycin polyketides is 3-amino-5-hydroxybenzoic acid **59** (AHBA) [94]. The biosynthesis of AHBA has recently been described by Floss and co-workers from the initial branch point of the shikimic acid pathway prior to 3-deoxy-D-*arabino*-heptulosonic acid 7-phosphate (DAHP) [95]. The pathway shown in Scheme 25 was delineated by feedings of the proposed AHBA precursors, in labelled forms, to cell-free extracts of both the rifamycin B producer *A. mediterranei* S699 and the ansatrienin A producer *S. collinus* Tü1892. In these experiments each of the compounds **61–64** was converted into AHBA with generally increasing efficiency. Most importantly the shikimate pathway compound DAHP cannot replace phosphoenolpyruvate **61** and erythrose 4-phosphate **60**, or aminoDAHP **62** as the precursor of AHBA **59**.

An mC_7N unit is also present in the core of the polyketide antibiotic asukamycin from *Streptomyces nodosus* subsp. *asukaensis* [96]. This has been shown to arise not from a variant of the shikimate pathway, but from the condensation of a C_4 unit from the TCA cycle, closely related to succinate, with a C_3 unit, possibly dihydroxyacetone phosphate, from the triose pool. Related studies concerning 3-amino-4-hydroxybenzoic acid biosynthesis in *Streptomyces murayamaensis* mutants MC2 and MC3 support this hypothesis [97].

The macrocyclisation or extra chain attachment to the mC_7N unit of these metabolites leading to the formation of an amide bond, is presumably catalysed by an amidase activity. Floss has reported the possible identification of a gene resposible for such an enzyme in the gene cluster for rifamycin B biosynthesis in *A. mediterranei* [98].

The ansamycin polyketide ansatrienin A **57** also contains a fully reduced cyclohexane carboxylic (CHC) unit attached to the macrocyclic structure via a

D-alanine residue. Other polyketides which contain such an arrangement are the trienomycins [99], while similar CHC chain terminating (synthase starter) units are found in a branch of asukamycin [96], and in the ω-cyclohexyl fatty acids of certain thermophilic bacteria [100]; substituted CHCs are also found as PKS starter units in the rapamycin family of polyketides (see Sect. 8). The cyclohexyl moieties of these compounds have been demonstrated to derive from the shikimate pathway.

Almost the entire pathway from shikimic acid **65** to CHC **70** has been delineated by experiments with the ansatrienin A producer *S. collinus* Tü 1892 [101]. All putative intermediates for this pathway were prepared in labelled form and fed to the producing organism and the **57** was isolated and was characterised by NMR spectroscopy. Based on these results the pathway shown in Scheme 26 was proposed. The initial step is a 1,4-conjugate elimination of the hydroxyl group at C-3 and a proton at C-6 which gives rise to a cross-conjugated dihydroxy diene **66**. This then undergoes reduction of the double bond conjugated with the carbonyl group. A further 1,4-conjugate elimination of the C-4 hydroxyl group and C-1 proton gives **67**. Reduction to **68** is followed by a further reduction and dehydration to **69**. The final steps involve isomerisation to bring the remaining double bond into conjugation and reduction to yield **70**. It is especially noteworthy that in this pathway the sequence of dehydrations and double bond reductions ensures that no intermediate is ever aromatic.

Comparable studies have been performed for the formation of ω-cyclohexyl fatty acids in *Alicyclobacillus acidocaldarius* and the pathways are identical [100]. A recent publication concerning this later pathway has shown that the final remaining stereochemical ambiguity, the stereochemistry of proton loss at C-6 in the initial 1,4-conjugate elimination of shikimate occurs with loss of the *pro*-6R proton [102]. This mirrors the stereochemistry of "normal" shikimate metabolism in the formation of chorismate from 5-enolpyruvylshikimate 3-phosphate.

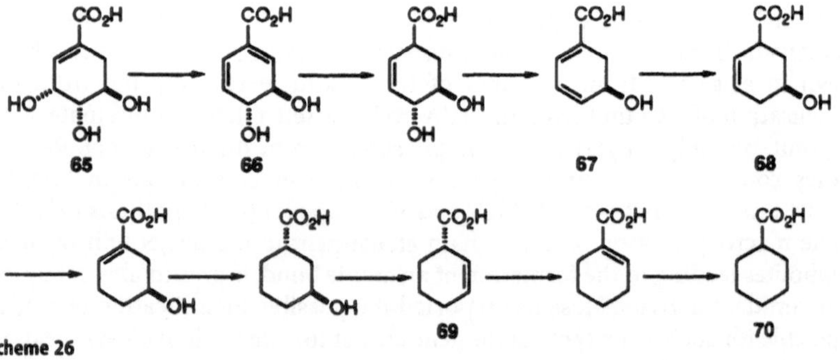

Scheme 26

8
Rapamycin, FK506 and FK520

The structurally related compounds rapamycin 71, FK506 72 and FK520 73 demonstrate antitumour, antifungal and immunosuppressant activities. This latter immunosuppressant activity has generated great interest due to applications in the therapeutic area of organ transplant surgery, and a detailed model for their mode of action has been developed [103]. Due to the similarity of their structures and the parallel biosynthetic studies, rapamycin will be the focus of this section and cross reference will be made where appropriate.

71

72 R = ⌇⟍⟋⟋

73 R = Et

Rapamycin obtained from *Streptomyces hygroscopicus* subsp. *hygroscopicus* is biosynthesised from seven acetate and seven propionate units, with O-methyl groups derived from L-methionine [104]. Competitive incorporation studies with radiolabelled precursors demonstrated that the pipecolate ring is derived from lysine via free pipecolate [105]. The *trans*-dihydroxycyclohexane carboxylic acid (DHCHC) ring is derived from shikimate [106], and recent studies have shown that free DHCHC can act as a precursor of the cyclohexyl unit [107]. Related experiments with *Streptomyces hygroscopicus* subsp. *yakushimaensis* have elaborated some of the steps involved in the formation of the cyclohexyl starter unit for FK520 [108]. The early steps closely resemble the biosynthesis of CHC in *S. collinus* as previously described. Shikimic acid undergoes a 1,4-conjugate elimination of the C-3 hydroxyl and C-6 hydrogen followed by reduction, double bond isomerisation and finally reduction to the saturated DHCHC.

The entire biosynthetic gene cluster for rapamycin biosynthesis has been sequenced and published [35]. The PKS genes were identified by hybridisation with DNA from the PKS genes for erythromycin biosynthesis. Sequencing beyond the PKS region identified other genes that are predicted to be involved in the late stages and regulation of rapamycin biosynthesis [109, 110]. The sequence and organisation of a gene encoding a four-module PKS protein which

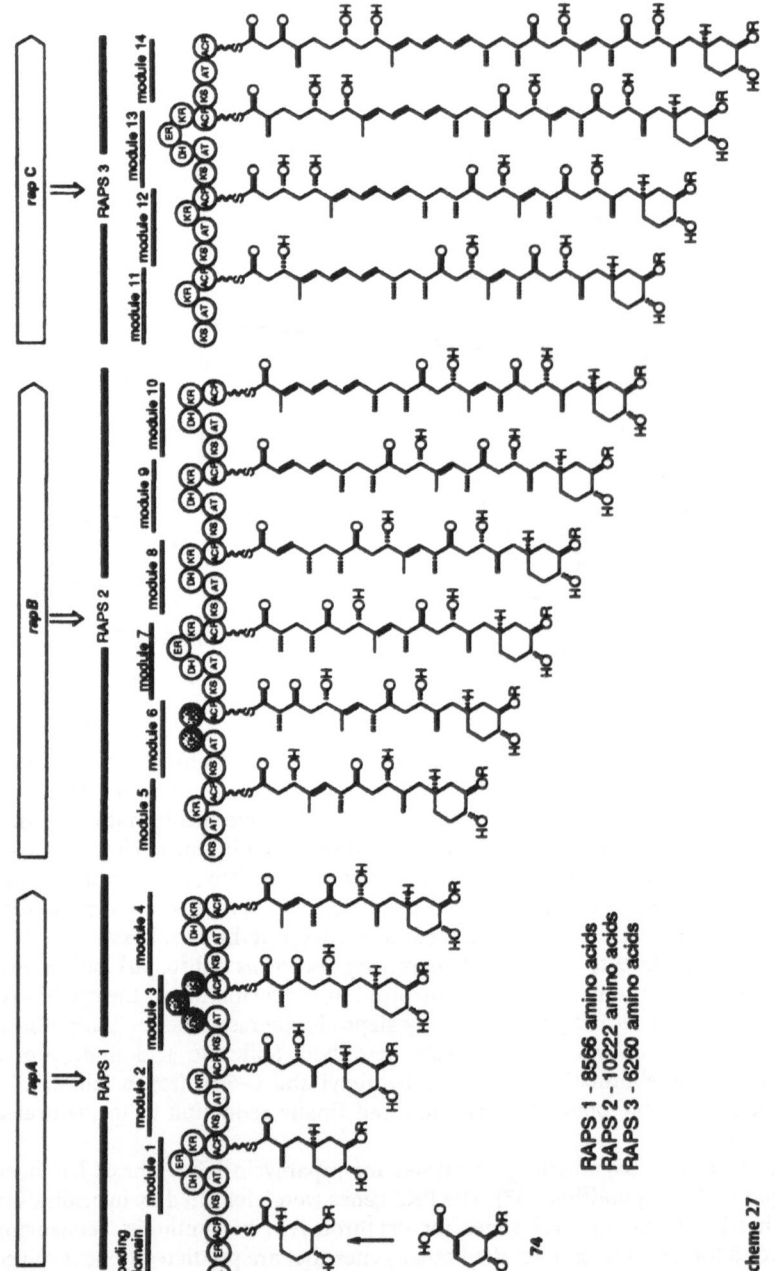

RAPS 1 - 8566 amino acids
RAPS 2 - 10222 amino acids
RAPS 3 - 6260 amino acids

Scheme 27

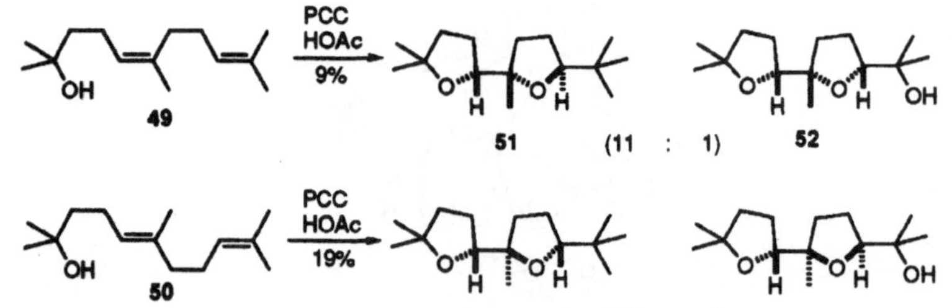

Scheme 23

bis-furans (Scheme 23) [84]. The formation of 51 and 52 from the (Z)-hydroxy-diene, and 53 and 54 from the (E)-hydroxydiene provide support for the syn-oxidative cyclisation mechanism. This work has been extended to ReO_3 reagents for the formation of a tris-furan system which is a core component of another polyether metabolite, goniafin [85].

6.2
Tetronasin

Tetronasin 55 is derived from seven acetate and six propionate units and a single L-methionine derived O-methyl group [86]. The origin of C-33 and C-34 remains obscure. Feeding experiments with NAC thioesters of di-, tri- and tetraketide chain elongation intermediates have been performed and the successful approach involved feeding to senescent (late) cells in which β-oxidative degradative pathways were minimal [87, 88]. It is worth noting at this point that in all current examples of feeding experiments with NAC thioesters, the proposed intermediate is only incorporated intact at the correctly predicted extension module, indicating some molecular recognition factor inherent to this process. This was extended in the study of tetronasin biosynthesis where all four possible diastereoisomers of the proposed tetraketide NAC intermediates were prepared and fed. Only the 'correct' diastereoisomer was incorporated intact [88]. These results are consistent with recent in vitro studies which demonstrate that the erythromycin diketide NAC intermediate acylates specifically the β-keto-synthase/acyl carrier protein domain of module 2 in DEBS 1 + TE [89].

In further studies several unnatural analogues of di-, tri- and tetraketide chain elongation intermediates have been incorporated into tetronasin analogues 55a-e (Scheme 24) [90]. These analogues contained either fluorine atoms in replacement of the α-proton or ethyl in place of methyl side chains, or a combination of the two. Isopropyl and benzyl side chains were not incorporated. The incorporation levels of 0.1-1.0% were significant, and those analogues which contained α-fluorine substitutions generally gave the highest incorporation levels. This substitution with fluorine has negligible stereochemical significance, and was designed to inhibit the α, β-dehydrogenation step of the β-oxidation degradative pathway.

A novel sodium ion templated polyene cyclisation reaction has been proposed for the biosynthesis of tetronasin, based on biomimetic studies [91]. This

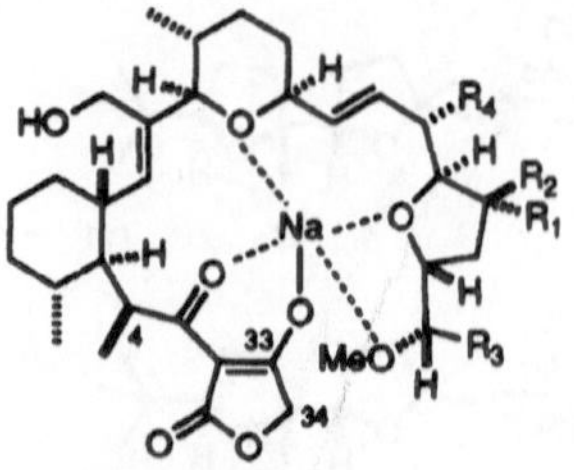

Scheme 24

55 R₁ = H, R₂ = R₃ = R₄ = Me (Tetronasin)

55a R₁ = F, R₂ = R₃ = R₄ = Me

55b R₁ = F, R₂ = Me, R₃ = Et, R₄ = Me

55c R₁ = H, R₂ = Et, R₃ = R₄ = Me

55d R₁ = F, R₂ = R₃ = Me, R₄ = Et

55e R₁ = H, R₂ = Me, R₃ = Et, R₄ = Me

biomimetic approach utilised a base-catalysed cascade which established two rings and four stereogenic centres in a model system closely related to a putative tetronasin precursor. The product obtained was different from the required structure only in the stereochemistry at C-4.

7
Ansamycin Antibiotics

The extensively studied rifamycin B 56 from *Amycolatopsis mediterranei* and ansatrienin A 57 from *Streptomyces collinus* exemplify this class of polyketide antibiotics. A common structural feature of these metabolites is the "ansa" bridge system of mixed (acetate/propionate) polyketide origin which is initiated from a "mC₇N" PKS starter unit. The presence of a (biosynthetically unique) mC₇N unit is recognised by a six-membered carbocyclic ring carrying the extra carbon and nitrogen atoms in a meta arrangement. This structure is generally quinonoid or benzenoid in nature, and found either with minimal further modification, as in 57, or as part of a naphthalenic structure, as in 56.

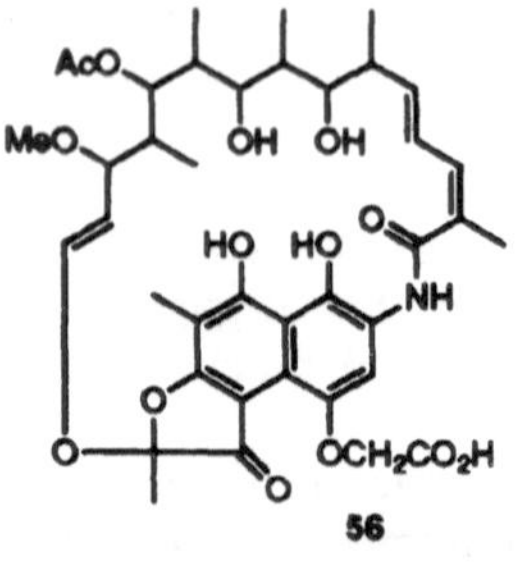

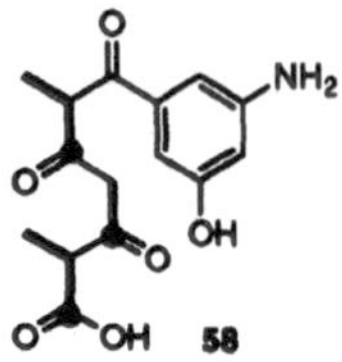

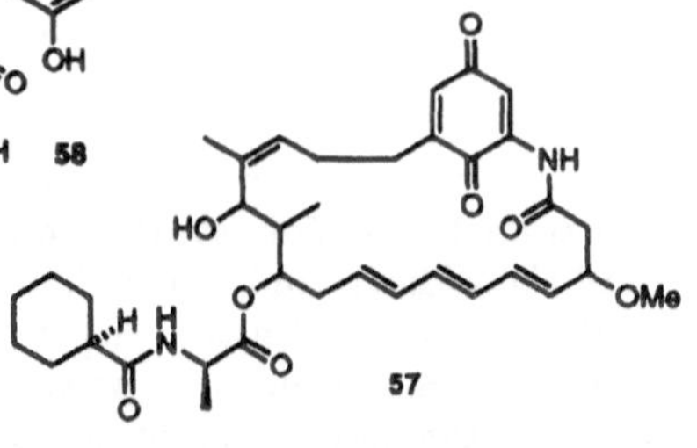

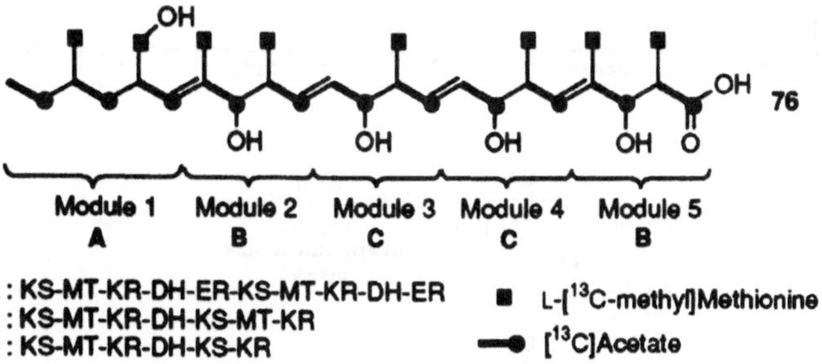

Fig. 8. Biosynthetic origin of the carbon atoms and proposed modular assembly of cubensic acid

ation, although the large number of individual recognition events required make this unlikely. Furthermore, many such methyl groups occur at reduced methylene positions for which an enzymic alkylation process utilising SAM is unreasonable.

A more satisfying proposal is that methyl transferase activities are present within the polyketide synthase, which catalyses the formation of carbon-carbon bonds at activated methylene groups of the β-keto thioester intermediates. This idea has been postulated by O'Hagan and co-workers, who have predicted a domain and module sequence of the putative PKS responsible for cubensic acid **76** biosynthesis (see Fig. 8) [115, 116].

Evidence for such a modular pathway has been provided from studies into the biosynthesis of the polyketide backbone **77** of $(4R)$-4-[(E)-2-butenyl]-4-methyl-L-threonine **78** which is incorporated into cyclosporin A in *Tolypocladium niveum* [117]. The proposed biosynthesis of **77** is presented in Scheme 29. In vitro studies using a cell extract have verified unambiguously that the biosynthetic mechanism is processive, that the first PKS free intermediate is the tetraketide **79**, and that methylation unequivocally occurs at the stage of the enzyme bound 3-oxo-4-hexenoic acid thioester **80** which is the triketide product from the second elongation cycle. These and other results indicate that the methyl transferase activity is inherent in the second module of the putative PKS.

At the genetic level preliminary evidence has been reported for a type I PKS implicated in the biosynthesis of the of the triol moiety of lovastatin (mevinolin) produced by *Aspergillus terreus*. This describes the presence of a methyltransferase domain within the PKS sequence [118].

9.3
Evidence for a Processive Mechanism: Feeding Studies

The biosynthesis of fungal aliphatic polyketides by a processive mechanism is further demonstrated by the successful incorporation of chain elongation intermediates as their NAC thioesters in several experiments [119, 120]. These

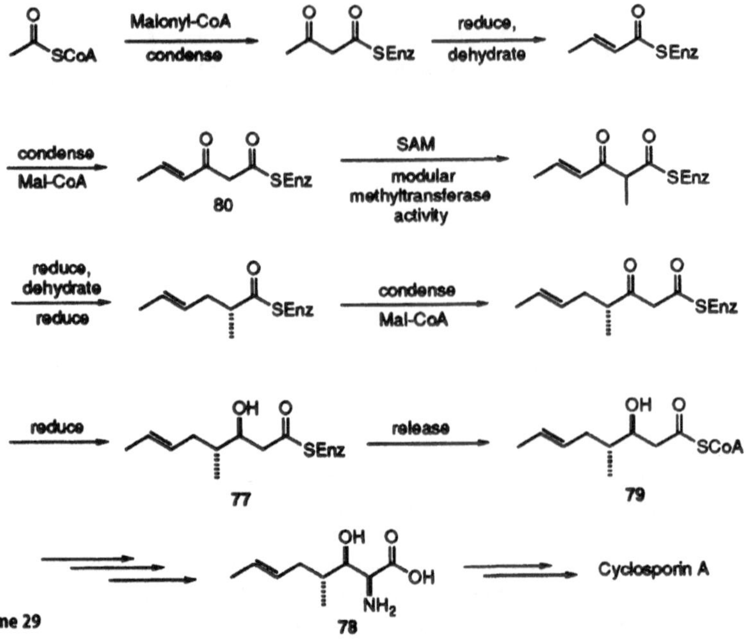

Scheme 29

experiments and the work of his own research group in this area are discussed in greater detail by Tom Simpson in Chap. 1.

9.4
The Epothilone Antibiotics

Worthy of a final note are the recently discovered epothilones. This family of compounds which has been isolated from several strains of *Sorangium cellulosum* [121] has attracted significant attention due to its potent biological activity and exciting mode of action [122]. Epothilones display narrow spectrum antimycotic activity against *Mucor hiemalis*, but are extremely cytotoxic towards animal cells. The mode of action is similar to that of taxol in stabilising microtubules during mitosis. In fact the epothilones are capable of displacing taxol from the requisite cellular binding site and are several orders of magnitude more active against multiply resistant cell lines. Epothilone B 81 which contains a methyl group attached to the epoxide ring has been shown to possess significantly greater activity than epothilone A 82.

While biosynthetic studies have yet to report on the epothilones, their structure is presumably elaborated from a polyacetate pathway with extra pendant methyl groups added from L-methionine; the structure also contains an interesting *gem* dimethyl grouping at C-4. The unusual thiazole starter unit can be predicted to arise either from cysteine directly with post-PKS addition of a C$_2$ unit (probably acetate) and cyclisation, or from an *N*-acetylcysteine, or from a

81 R = CH₃, Epothilone B
82 R = H, Epothilone A

starting thiazole structure prederived from cysteine and acetate. The impressive cytotoxicity of these structures has given rise to a flurry of interest in their total synthesis which was first performed for both **82** and **81** by Danishefsky and co-workers [123, 124].

10
References

1. Celmer WD (1965) J Am Chem Soc 87:1801
2. Omura S (ed)(1984) Macrolide antibiotics:chemistry, biology and practice. Academic Press, New York
3. McGuire JM, Bunch RL, Anderson RC, Boaz HE, Flynn EH, Powell M, Smith JW (1952) Antibiotics and Chemotherapy 2:281
4. Wiley PF, Gerzon K, Flynn EH, Sigal MV Jr, Weaver O, Quarck UC, Chauvette RR, Monahan R (1957) J Am Chem Soc 79:6062
5. Harris DR, McGeachin SG, Mills HH (1965) Tetrahedron Lett 6:679
6. Martin JR, Rosenbrook W (1967) Biochemistry 6:435
7. Corcoran JW, Vygantas AM (1982) Biochemistry 21:263
8. Weber JM, Wierman CK, Hutchinson CR (1985) J Bacteriol 164:425
9. Corcoran JW (1981). In:Corcoran JW (ed) Antibiotics, vol IV: biosynthesis. Springer, Berlin Heidelberg New York, p 132
10. Vygantas AM, Corcoran JW (1974) Fed Proc 33:1233
11. Hung PP, Marks CL, Tardrew PL (1965) J Biol Chem 240:1322
12. Martin JR, Perun TJ, Girolami RL (1966) Biochemistry 5:2852
13. Weber JM, Leung JO, Maine GT, Potenz RHB, Paulus TJ, DeWitt JP (1990) J Bacteriol 172:2372
14. Majer J, Martin JR, Egan RS, Corcoran JW (1977) J Am Chem Soc 99:1620
15. Corcoran JW, Vygantas AM (1977) Fed Proc 36:663
16. Corcoran JW (1975) In:Hash JH (ed) Methods in enzymology XLIII. Academic Press, London, chap. 33, p 487
17. Lambalot RH, Cane DE, Aparicio JJ, Katz L (1995) Biochemistry 34:1858
18. Katz L, Donadio S (1995). In:Vining LC, Stoddard C (eds) Genetics and biochemistry of antibiotic production. Butterworth-Heinmann, Boston, p 223
19. Liu H-W, Thorson JS (1994) Annu Rev Microbiol 48:223
20. Kaneda T, Butte JC, Taubman SB, Corcoran JW (1962) J Biol Chem 237:322
21. Cane DE, Hasler H, Liang T (1981) J Am Chem Soc 103:5960
22. Cane DE, Yang C-CJ (1987) J Am Chem Soc 109:1255
23. Cane DE, Prabhakaran PC, Tan W, Ott WR (1991) Tetrahedron Lett 32:5457
24. Yue S, Duncan JS, Yamamoto Y, Hutchinson CR (1987) J Am Chem Soc 109:1253
25. Cane DE, Lambalot RH, Prabhakaran PC, Ott WR (1993) J Am Chem Soc 115:522
26. Wakil S J (1989) Biochemistry 28:4523

27. Smith S (1994) FASEB 8:1248
28. Fulco AJ (1983) Prog Lipid Res 22:133
29. Hopwood DA, Sherman DH (1990) Annu Rev Genet 24:37
30. Cortés J, Haydock SF, Roberts GA, Bevitt DJ, Leadlay PF (1990) Nature 348:176
31. Donadio S, Staver MJ, McAlpine JB, Swanson SJ, Katz L (1991) Science 252:675
32. Donadio S, Katz L (1992) Gene 111:51
33. Bevitt DJ, Cortes J, Haydock SF, Leadlay PF (1992) Eur J Biochem 204:39
34. Staunton J (1991) Angew Chem Int Ed Engl 30:1302
35. Schwecke T, Aparicio JF, Molnár I, König A, Khaw LE, Haydock SF, Oliynyk M, Caffrey P, Cortés J, Lester JB, Böhm GA, Staunton J, Leadlay PF (1995) Proc Natl Acad Sci USA 92:7839
36. Kuhstoss S, Huber M, Turner JR, Pashal JW, Rao RN (1996) Gene 183:231
37. Donadio S, Staver MJ, McAlpine JB, Swanson SJ, Katz L (1991) Science 252:675
38. Donadio S, McAlpine JB, Sheldon PL, Jackson M, Katz L (1993) Proc Natl Acad Sci USA 90:7119
39. Caffrey P, Bevitt DJ, Staunton J, Leadlay PF (1992) FEBS Lett 304:225
40. Staunton J, Caffrey P, Aparicio JF, Roberts GA, Bethell SS, Leadlay PF (1996) Nature Struct Biol 3:188
41. Aparicio JF, Caffrey P, Marsden AFA, Staunton J, Leadlay PF (1994) J Biol Chem 269:8524
42. Marsden AFA, Caffrey P, Aparicio JF, Loughran MS, Staunton J, Leadlay PF (1994) Science 263:378
43. Kao CM, Pieper R, Cane DE, Khosla C (1996) Biochemistry 35:12363
44. McDaniel R, Ebert-Khosla S, Hopwood DA, Khosla C (1993) Science 262:1546
45. Kao CM, Katz L, Khosla C (1994) Science 265:509
46. Cane DE, Luo G, Khosla C, Kao CM, Katz L (1995) J Antibiot 48:647
47. Cortés J, Wiesmann KEH, Roberts GA, Brown MJB, Staunton J, Leadlay PF (1995) Science 268:1487
48. Kao CM, Luo G, Katz L, Cane DE, Khosla C (1995) J Am Chem Soc 117:9105
49. Kao CM, Luo G, Katz L, Cane DE, Khosla C (1996) J Am Chem Soc 118:9184
50. Oliynyk M, Brown MJB, Cortés J, Staunton J, Leadlay PF (1996) Chem Biol 3:833
51. Leadlay P F, Staunton J, Marsden AFA, Wilkinson B, Dunster NJ, Cortés J, Oliynyk M, Hanefeld U, Brown MJB (1997) In:Baltz RH, Hegeman GD, Skatrud PL (eds) Industrial micro-organisms:basic and applied molecular genetics. American Society for Microbiology, Washington DC, in press
52. Wiesmann KEH, Cortés J, Brown MJB, Cutter AL, Staunton J, Leadlay PF (1995) Chem Biol 2:583
53. Pieper R, Luo G, Cane DE, Khosla C (1995) J Am Chem Soc 117:11373
54. Luo G, Pieper R, Rosa A, Khosla C, Cane DE (1996) Bioorg Med Chem Lett 4:995
55. O'Hagan D (1991) The polyketide metabolites. Ellis Horwood, Chichester, p 70
56. Pieper R, Luo G, Cane DE, Khosla C (1995) Nature 378:263
57. Pieper R, Gokhale RS, Luo G, Cane DE, Khosla C (1997) Biochemistry 36:1846
58. Caffrey P, Green B, Packman LC, Rawlings BJ, Staunton J, Leadlay PF (1991) Eur J Biochem 195:823
59. Aggarwal R, Caffrey P, Leadlay PF, Smith CJ, Staunton J (1995) J Chem Soc Chem Commun:1519
60. Burg RW, Miller BM, Baker EE, Birnbaum J, Currie SA, Hartman R, Kong Y-L, Monaghan RL, Olson G, Putter I, Tunac JB, Wallick H, Stapley EO, Oiwa R, Omura S (1979) Antimicrob Agents Chemother 15:361
61. (a) Chabala JC, Mrozik H, Tolman RL, Eskola P, Lusi A, Peterson LH, Woods MF, Fisher MH (1980) J Med Chem 29:1134; (b) Remme J, Baker RHA, DeSole G, Dadzie KY, Walsh JF, Adams MA, Alley ES, Avissey HSK (1989) Trop Med Parasit 40:367
62. Cane DE, Liang T-C, Kaplan LK, Nallin MK, Schulman MD, Hensens OD, Douglas AW, Albers-Schönberg G (1983) J Am Chem Soc 105:4110
63. Chen ST, Hensens OD, Schulman MD (1989). In: Campbell WC (ed) Ivermectin and abamectin. Springer, Berlin Heidelberg New York, pp 55–72

64. Chen ST, Arison BH, Gullo VP, Inamine ESJ (1989) Indust Microbiol 4:231
65. MacNeil DJ, Occi JL, Gewain KM, MacNeil T, Gibbons PH, Ruby CL, Danis SJ (1992) Gene 115:119 and references cited therein
66. (a) Dutton CJ, Gibson SP, Goudie AC, Holdom KS, Pacey MS, Ruddock JC, Bu'Lock JD, Richards MK (1991) J Antibiot 44:357; (b) Hafner EW, Holley BW, Holdom KS, Lee SE, Wax RG, Beck D, McArthur HA, Wernau WC (1991) J Antibiot 44:349
67. Dutton CJ, Hooper AM, Leadlay PF, Staunton J (1994) Tetrahedron Lett 35:327
68. Omura S, Takeshima H, Nakagawa A, Miyazawa J, Piriou F, Lukacs G (1977) Biochemistry 16:2860
69. O'Hagan D, Robinson JA, Turner DL (1983) J Chem Soc Chem Commun:1337 and references cited therein
70. Huber MLB, Paschal JW, Leeds JP, Hirst HA, Wind JA, Miller FD, Turner JR (1990) Antimicrob Agents Chemother 34:1535
71. Baltz RH, Seno ET (1988) Ann Rev Microbiol 42:547
72. Cane DE, Tan WT, Ott WR (1993) J Am Chem Soc 115:527 and references cited therein
73. Cane DE, Luo G (1995) J Am Chem Soc 117:6633
74. Oikawa H, Suzuki Y, Naya A, Katayama K, Ichihara A (1994) J Am Chem Soc 116:3605
75. Oikawa H, Katayama K, Suzuki Y, Ichihara A (1995) J Chem Soc Chem Commun:1321
76. Reynolds K, Robinson JA (1985) J Chem Soc Chem Commun:1831
77. Patzelt H, Robinson JA (1993) J Chem Soc Chem Commun:1258
78. Westley JW, Blount JF, Evans RH Jr, Stempel A, Berger J (1974) J Antibiot 27:597
79. Cane DE, Celmer WD, Westley JW (1983) J Am Chem Soc 105:3594
80. (a) Cane DE, Liang T-C, Hasler H (1982) J Am Chem Soc 104:7274; (b) Ajaz AA, Robinson JA, Turner DL (1987) J Chem Soc Perkin Trans 1:27
81. (a) Paterson I, Tillyer RD, Smaill JB (1993) Tetrahedron Lett 34:7137; (b) Russell SA, Robinson JA, Williams DJ (1987) J Chem Soc Chem Commun:351 and references cited therein
82. Holmes DS, Sherringham JA, Dyer UC, Russell ST, Robinson JA (1990) Helv Chim Acta 73:239
83. Townsend CA, Basak A (1991) Tetrahedron 47:2591
84. McDonald FE, Towne TB (1994) J Am Chem Soc 116:7921
85. McDonald FE, Towne TB, Schultz CC (1996) Lecture and Abstract at the 20th IUPAC Symposium on the Chemistry of Natural Products' Chicago Illinois USA, 15–20 September, Abstract SP-39
86. (a) Bulsing JM, Laue ED, Leeper FJ, Staunton J, Davies DH, Ritchie GAF, Davies A, Davies AB, Mabelis RP (1984) J Chem Soc Chem Commun:1301; (b) Demetriadou AK, Laue ED, Staunton J, Ritchie GAF, Davies A, Davies AB (1985) J Chem Soc Chem Commun:408
87. Hailes HC, Jackson CM, Leadlay PF, Ley SV, Staunton J (1994) Tetrahedron Lett 35:307
88. Hailes HC, Handa S, Leadlay PF, Lennon IC, Ley SV, Staunton J (1994) Tetrahedron Lett 35:311
89. Tsukamoto N, Chuck J-A, Luo G, Kao CM, Khosla C, Cane DE (1996) Biochemistry 35:15244
90. Less SL, Handa S, Leadlay PF, Dutton CJ, Staunton J (1996) Tetrahedron Lett 37:3511
91. Boons G-J, Lennon IC, Ley SV, Owen ESE, Staunton J, Wadsworth DJ (1994) Tetrahedron Lett 35:323
92. Ghisalba O, Fuhrer H, Richter WJ, Moss S (1981) J Antibiot 34:58
93. White RJ, Martinelli E, Gallo GG, Lancini G, Beynon P (1973) Nature 243:273
94. (a) Kibby JJ, McDonald IA, Rickards RW (1980) J Chem Soc Chem Commun 768; (b) Ghisalba O, Nüesch J (1981) J Antibiot 34:64
95. Kim C-G, Kirschning A, Bergon P, Zhou P, Su E, Sauerbrei B, Ning S, Ahn Y, Breuer M, Leistner E, Floss HG (1996) J Am Chem Soc 118:7486
96. Thiericke R, Zeeck A, Akira N, Omura S, Herrold RE, Wu STS, Beale JM, Floss HG (1990) J Am Chem Soc 112:3979
97. Gould SJ, Melville CR, Cone MC (1996) J Am Chem Soc 118:9228

98. Floss HG (1996) Lecture at the Royal Society of Chemistry Perkin Division and Bio-organic Group International interdisciplinary Meeting on 'Polyketides:Chemistry Bio-chemistry and Genetics', University of Bristol 1–3 April 1996
99. Funayama S, Okada K, Iwasaki K, Komiyama K, Umezawa I (1985) J Antibiot 38:1677 and references cited therein
100. Moore BS, Poralla K, Floss HG (1993) J Am Chem Soc 115:5267
101. Moore BS, Cho H, Casati R, Kennedy E, Reynolds KA, Mocek U, Beale JM, Floss HG (1993) J Am Chem Soc 115:5254
102. Handa S, Floss HG (1997) J Chem Soc Chem Commun:153
103. (a) Schreiber SL, Crabtree GR (1992) Immunol Today 13:136; (b) Kunz J, Hall MN (1993) Trends Biochem Sci 18:334; (c) Dumont FJ, Su Q (1995) Life Sciences 58:373
104. Paiva NL, Demain AL, Roberts MF (1991) J Nat Prod 54:167
105. Paiva NL, Demain AL, Roberts MF (1993) Enzyme Microb Technol 15:581
106. Paiva NL, Roberts MF, Demain AL (1993) J Ind Microbiol 12:423
107. Lowden PAS, Böhm GA, Staunton J, Leadlay PF (1996) Angew Chem Int Ed Engl 35:2249
108. Wallace KW, Reynolds KA, Koch K, McArthur HAI, Brown MS, Wax RG, Moore BS (1994) J Am Chem Soc 116:11600
109. Molnár I, Aparicio JF, Haydock SF, Khaw LE, Schwecke T, König A, Staunton J, Leadlay PF (1996) Gene 169:1
110. Aparicio JF, Molnár I, Schwecke T, König A, Haydock SF, Khaw LE, Staunton J, Leadlay PF (1996) Gene 169:9
111. Motamedi H, Cai S-J, Shafiee A, Elliston KO (1997) Eur J Biochem 244:74
112. KleinKauf H, Von Döhren H (1996) Eur J Biochem 236:335
113. König A, Schwecke T, Molnár I, Böhm GA, Lowden PAS, Staunton J, Leadlay PF (1997) Eur J Biochem 247:526
114. Nielsen JB, Hsu M-J, Byrne KM, Kaplan L (1991) Biochemistry 30:5789
115. O'Hagan D, Rogers SV, Duffin GR, Edwards RL (1992) Tetrahedron Lett 33:5585
116. O'Hagan D (1991) The polyketide metabolites. Ellis Horwood, Chichester, pp 86–90
117. Offenzeller M, Santer G, Totschnig K, Su Z, Moser H, Traber R, Schneider-Scherzer E (1996) Biochemistry 35:8401
118. Davis R, Aldrich TL, Nguyen DK, Hendrickson LE, Roach C, Vinci VA, McAda PC (1994) Poster and Abstract at the '7th International Symposium on the Genetics of Industrial Microorganisms' Palais des Congrés Montréal Québec Canada 26 June-1 July 1994, Programme and Abstracts p 288, p 192
119. Jacobs A, Staunton J (1995) J Chem Soc Chem Commun:863 and references cited therein
120. Li Z, Martin M, Vederas J (1992) J Am Chem Soc 114:1531
121. Gerth K, Bedorf N, Höfle G, Irschik H, Reichenbach H (1996) J Antibiot 49:560 and references cited therein
122. Bollag DM, McQueney PA, Zhu J, Hensens O, Koupal L, Liesch J, Goetz M, Lazarides E, Woods CM (1995) Cancer Res 55:2325
123. Balog A, Meng D, Kamenecka T, Bertinato P, Su D-S, Sorensen EJ, Danishefsky SJ (1996) Angew Chem Int Ed Engl 35:2801
124. Su D-S, Meng D, Bertinato P, Balog A, Sorensen EJ, Danishefsky SJ, Zheng Y-H, Chou T-C, He L, Horwitz SB (1997) Angew Chem Int Ed Engl 36:757

Cofactor Biosynthesis: A Mechanistic Perspective

Tadhg P. Begley*· Cynthia Kinsland · Sean Taylor ·
Manish Tandon · Robb Nicewonger · Min Wu · Hsiu-Ju Chiu · Neil Kelleher ·
Nino Campobasso · Yi Zhang

* Department of Chemistry, Cornell University, Ithaca, NY 14853, USA.
E-mail tpb2@cornell.edu

The chemistry of the cofactors has provided a fertile area of overlap between organic chemi-stry and biochemistry, and the organic chemistry of the cofactors is now a thoroughly studied area. In contrast, the chemistry of cofactor biosynthesis is still relatively underdeveloped. In this review the biosynthesis of nicotinamide adenine dinucleotide, riboflavin, folate, molyb-dopterin, thiamin, biotin, lipoic acid, pantothenic acid, coenzyme A, S-adenosylmethionine, pyridoxal phosphate, ubiquinone and menaquinone in *E. coli* will be described with a focus on unsolved mechanistic problems.

Keywords: Cofactor, vitamin, *E. coli*, biosynthesis, mechanism.

1
Introduction

The range of functionality provided by the 20 amino acids found in proteins consists of weak acids and bases, nucleophiles, hydrogen bond donors and acceptors, and the redox active thiol/disulfide. This limited range of chemistry is inadequate for the catalysis of many reactions found to occur in biological systems. Therefore, a variety of small organic molecules, called cofactors, coenzymes, or vitamins, have evolved to broaden the limited range of chemistry that can be catalyzed by simple proteins.

The chemistry of the cofactors has provided a fertile area of overlap between organic chemistry and biochemistry, and the organic chemistry of the cofactors is now a thoroughly studied area. In contrast, the chemistry of cofactor biosynthesis is still relatively underdeveloped. In this review the biosynthesis of the cofactors shown in Fig. 1 will be described. Heme and cobalamin will be omitted as these are covered in the review by Allan Battersby and Finian Leeper in this volume. We will focus on cofactor biosynthesis in *Escherichia coli* because the relevant genetics and biochemistry have been most intensively studied in this organism and it is the easiest system in which to carry out molecular biology [1]. Enzymes from other sources will be described only if the corresponding enzyme from *E. coli* has not been isolated or subjected to mechanistic studies.

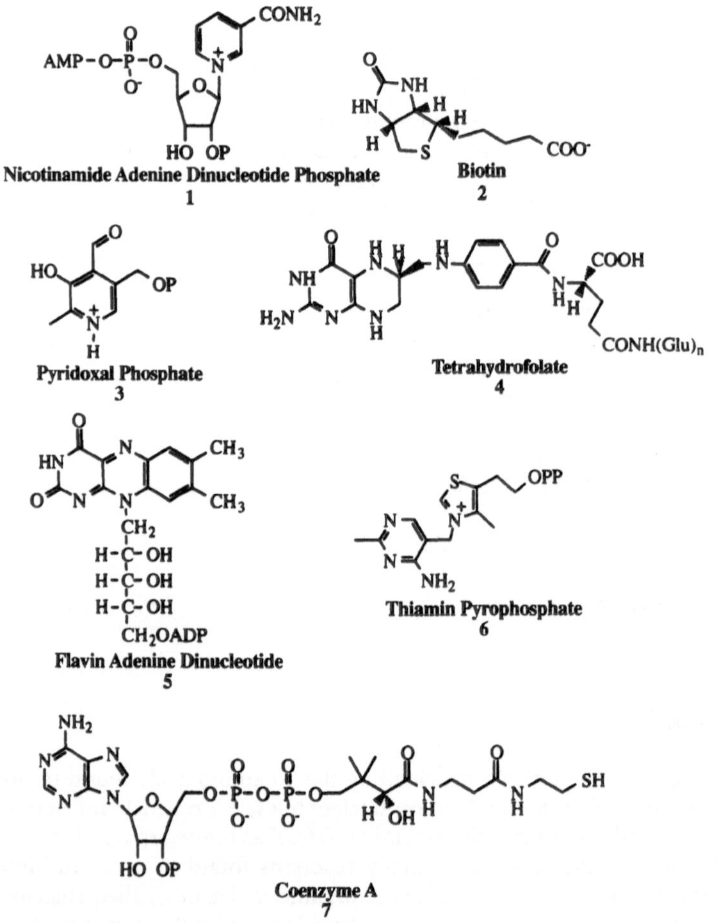

Fig. 1. The cofactors covered in this review

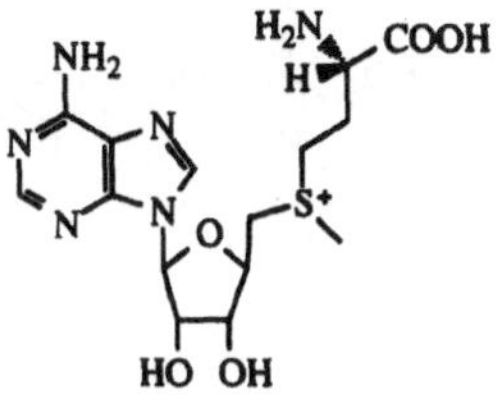

H_2N COOH

S-Adenosyl Methionine
8

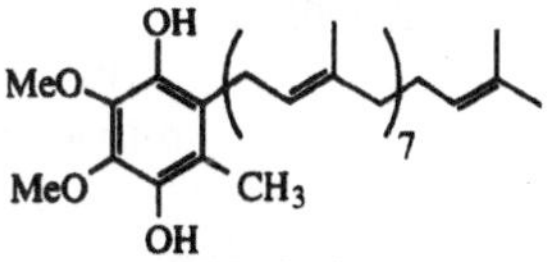

Ubiquinol
9

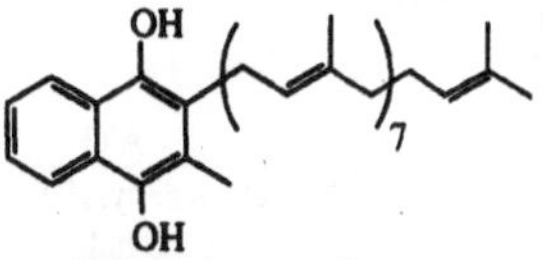

Dihydromenaquinone
10

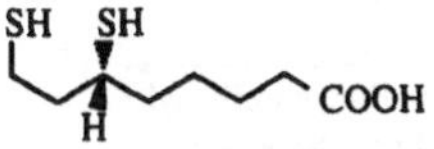

Lipoic Acid
11

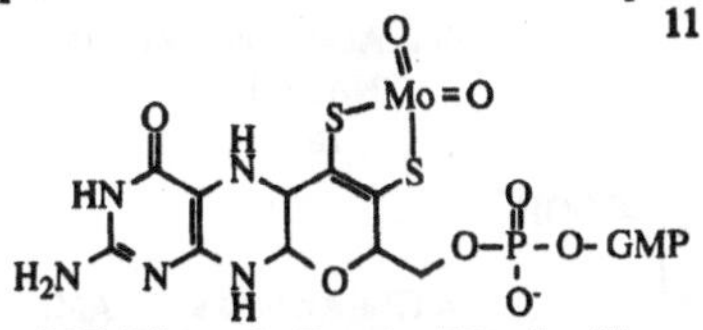

Molybdopterin Guanine Dinucleotide
12

Fig. 1 (continued)

Even though *E. coli* is a very well-studied bacterium, many interesting mechanistic problems in cofactor biosynthesis in this organism remain unsolved. The mechanisms for the formation of the nicotinamide ring of NAD^+, the pyridine ring of pyridoxal, the pterin system of molybdopterin, and the thiazole and pyrimidine rings of thiamin are unknown. The sulfur transfer chemistry involved in the biosynthesis of lipoic acid, biotin, thiamin and molybdopterin is not yet understood. The formation of the isopentenylpyrophosphate precursor to the prenyl side chain of ubiquinone and menaquinone does not occur by the mevalonate pathway. None of the enzymes involved in this alternative terpene biosynthetic pathway have been characterized. The aim of this review is to focus attention on these unsolved mechanistic problems.

2
Nicotinamide Adenine Dinucleotide Biosynthesis

Nicotinamide adenine dinucleotide (NAD^+ and $NADP^+$) is the biochemical hydride donor/acceptor [2].

2.1
Biosynthetic Pathway

The biosynthesis of $NAD(P)^+$ in *E. coli* is outlined in Fig. 2 [3]. Oxidation of aspartic acid to the imine **14** followed by condensation with dihydroxyacetone

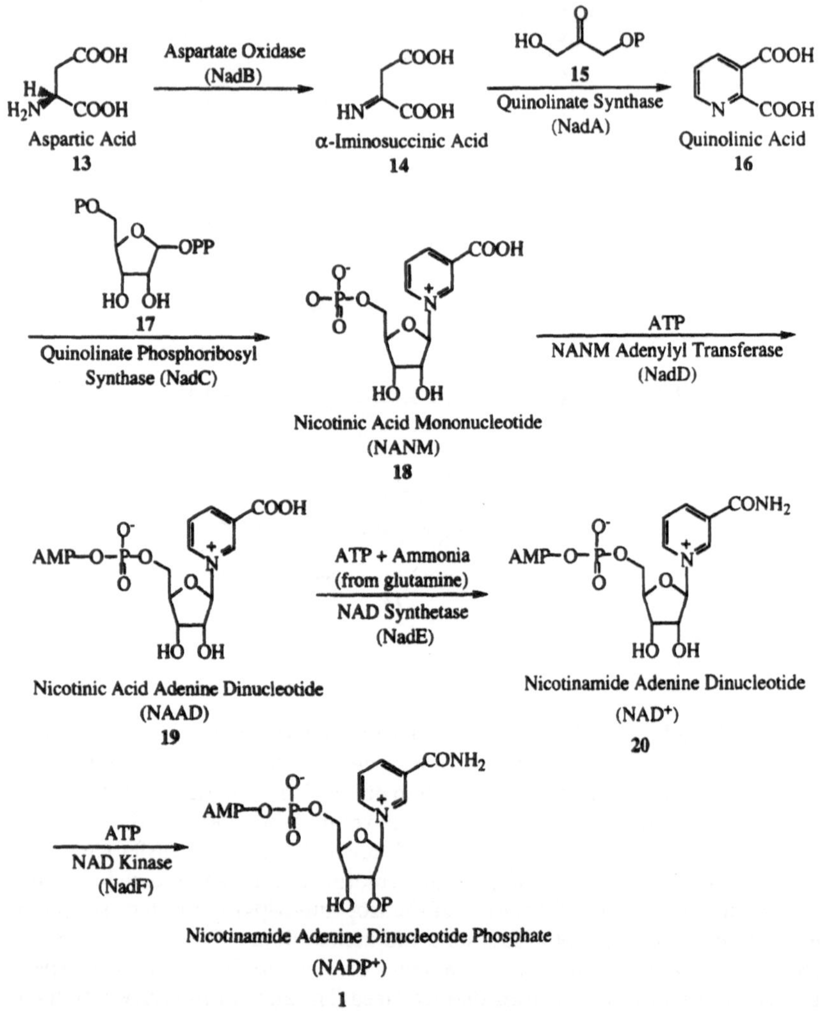

Fig. 2. The nicotinamide adenine dinucleotide biosynthetic pathway

phosphate gives quinolinic acid. Ribosylation, followed by decarboxylation, adenylation and amide formation, completes the biosynthesis of NAD⁺. An additional phosphorylation gives NADP⁺.

2.2
Enzyme Overexpression and Purification

The current status of the overexpression and purification of the nicotinamide adenine dinucleotide biosynthetic enzymes is summarized in the table [3].

Gene	Sequenced?	Overexpressed?	Purified?	Other source [4]	Reference
nadB	yes	yes	yes		5–7
nadA	yes	yes	no	no	6, 8
nadC	yes	yes	yes		9
nadD	no	no	no	no	3
nadE	yes	yes	yes		10, 11
nadF	no	no	no	no	12

2.3
Mechanistic Highlights

2.3.1
Aspartate Oxidase

A mechanistic proposal for this flavoenzyme, analogous to the mechanism of monoamine oxidase [13], is outlined in Fig. 3. No mechanistic studies have been described.

2.3.2
Quinolinate Synthase

This enzyme has been overexpressed but not purified due to instability. The mechanism for the formation of quinolinic acid is unknown. A proposal is outlined in Fig. 4.

2.3.3
Quinolinate Phosphoribosyl Transferase

The mechanism for this reaction has not been studied. A proposal is outlined in Fig. 5 [9].

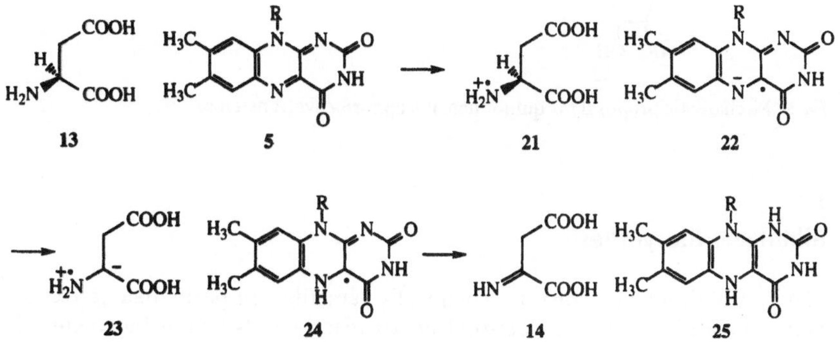

Fig. 3. Mechanistic proposal for aspartate oxidase

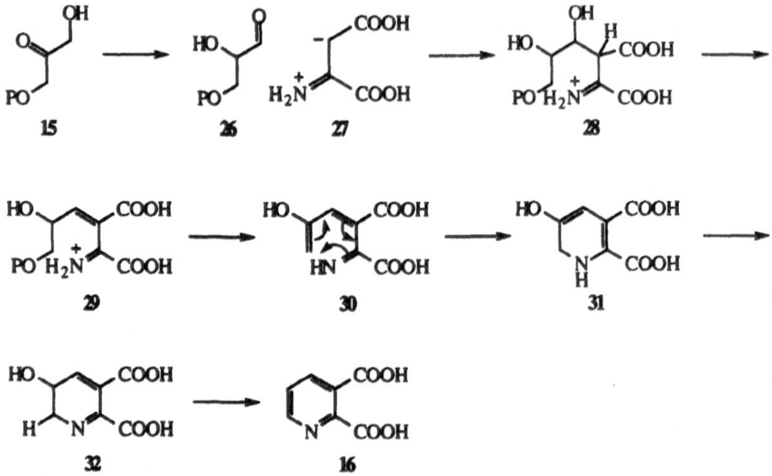

Fig. 4. Mechanistic proposal for quinolinate synthase

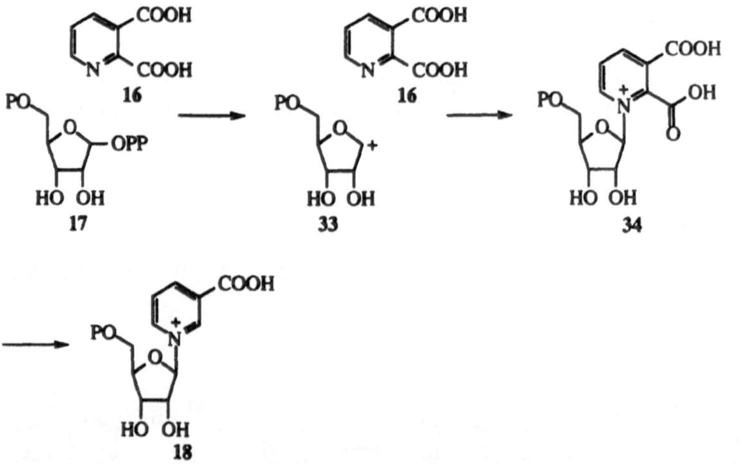

Fig. 5. Mechanistic proposal for quinolinate phsophoribosyl transferase

3
Riboflavin Biosynthesis

Flavin containing cofactors are chemically versatile and participate in oxygen activation and in 1- and 2-electron transfer reactions [14]. In its biochemically active forms, flavin occurs as flavin mononucleotide (FMN) and flavin adenine dinucleotide (FAD).

3.1
Biosynthetic Pathway

The biosynthesis of FAD is summarized in Fig. 6 [15, 16]. Loss of C-8 and pyrophosphate from guanosine triphosphate gives **36**. Deamination of the pyrimidine, ring opening of the ribose, reduction, and dephosphorylation gives **39**. Condensation of this with **40** gives lumazine **41**. Two molecules of lumazine undergo a remarkable coupling reaction to give riboflavin. Phosphorylation followed by adenylation gives FAD.

3.2
Enzyme Overexpression and Purification

The current status of the overexpression and purification of the riboflavin biosynthetic enzymes is summarized in the table [15, 17, 18].

Gene	Sequenced?	Overexpressed?	Purified?	Other source [4]	Reference
ribA	yes	yes	yes		19
ribD	yes	yes	yes		20
ribB	yes	yes	yes		21
ribE	yes	yes	yes		22, 23
ribC	yes	yes	yes	yes	24
ribF	yes	no	no	yes	15, 25

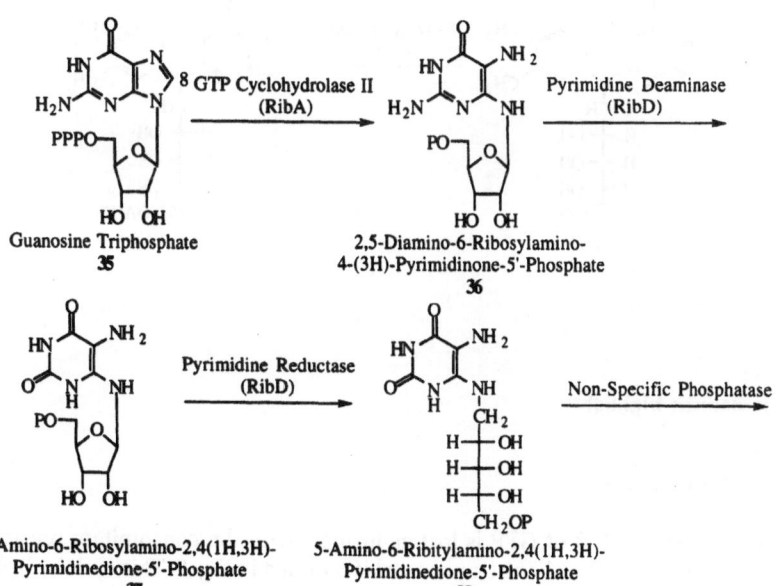

Fig. 6. The flavin adenine dinucleotide biosynthetic pathway

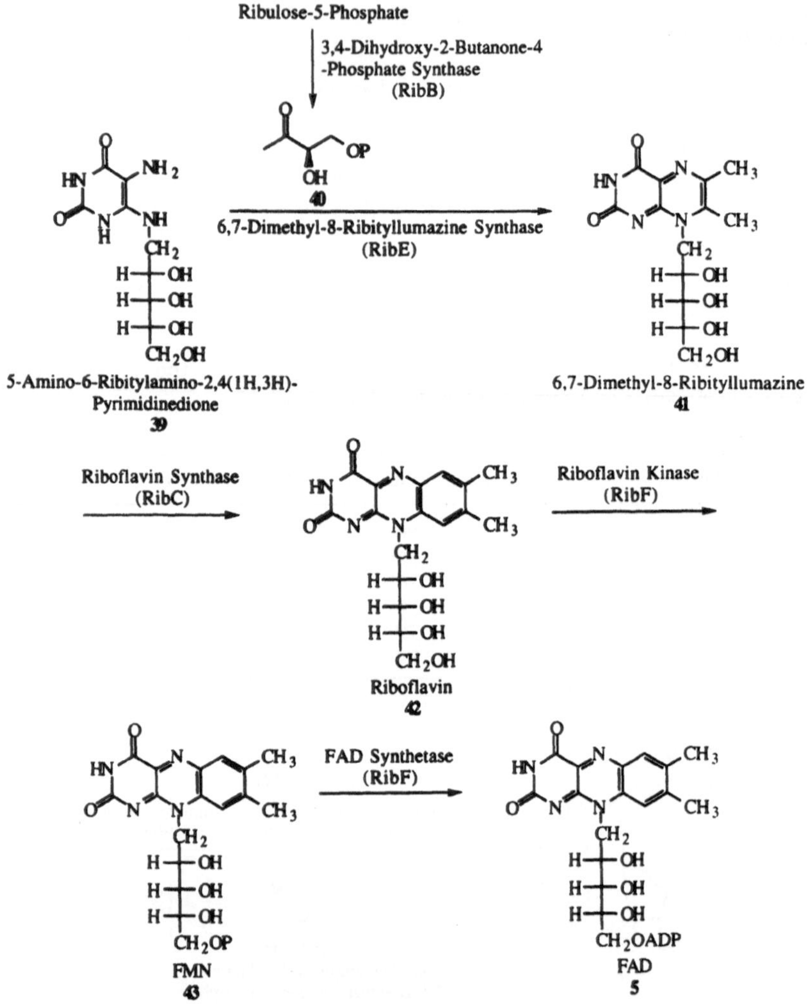

Fig. 6 (continued)

3.3
Mechanistic Highlights

3.3.1
GTP Cyclohydrolase II

In this reaction C-8 of GTP is lost as formate with the concomitant release of pyrophosphate. A mechanistic proposal is outlined in Fig. 7. This cyclohydrolase is different from the GTP cyclohydrolase I involved in folate biosynthesis (see section 4 of this review).

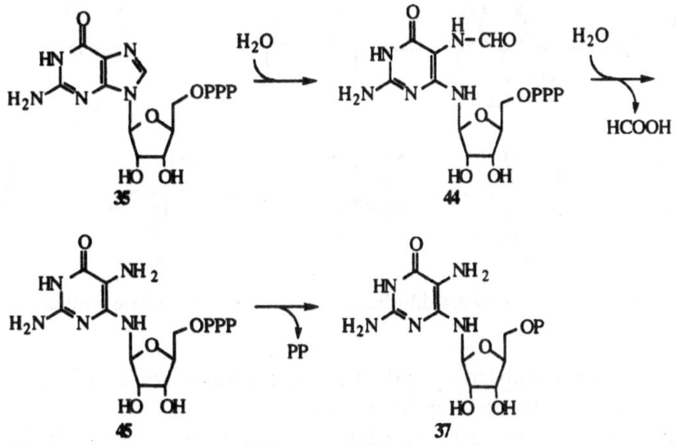

Fig. 7. Mechanistic proposal for GTP cyclohydrolase II

3.3.2
Pyrimidine Reductase

A mechanistic proposal for this reaction is outlined in Fig. 8. The pro-S proton at C-1 of **38** is derived from nicotinamide. This excludes an alternate mechanism involving an Amadori rearrangement [26].

3.3.3
3,4-Dihydroxy-2-butanone-4-phosphate Synthase

A postulated mechanism for the conversion of **47** to **40**, which is consistent with extensive labeling studies carried out on the enzyme isolated from *Candida guilliermondi,* is outlined in Fig. 9 [27, 28, 29, 30].

3.3.4
6,7-Dimethyl-8-ribityl Lumazine Synthase

Mechanistic [31] and structural [32] studies have been carried out on the enzyme isolated from *B. subtilis,* which exists as a large $\alpha_3\beta_{60}$ complex. The β-subunits catalyze the lumazine synthase reaction and are arranged as an icosahedral shell

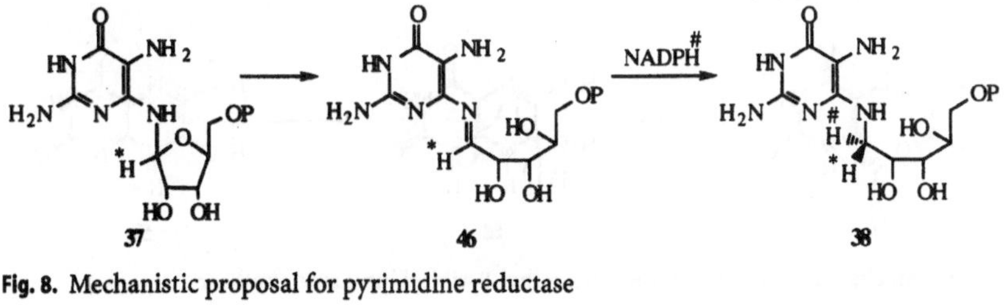

Fig. 8. Mechanistic proposal for pyrimidine reductase

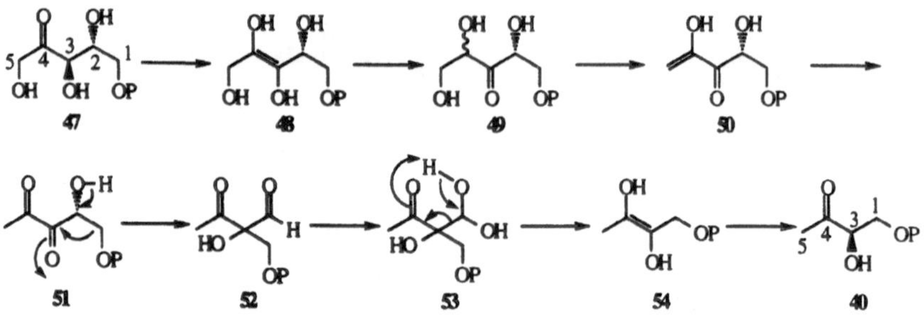

Fig. 9. Mechanistic proposal for 3,4-dihydroxy-2-butanone-4-phosphate synthase

surrounding a core consisting of the three α-subunits. The α-subunits catalyze the synthesis of riboflavin from lumazine.

A mechanistic proposal for lumazine synthase, consistent with the crystal structure and with the observation that the methyl group of **40** ends up as the C-6 methyl group of lumazine, is outlined in Fig. 10 [31, 33]. Butanedione is not an intermediate and none of the proposed intermediates have yet been trapped.

3.3.5
Riboflavin Synthase

Riboflavin synthase catalyzes the disproportionation of two molecules of lumazine to give riboflavin and **39** [34]. This complex reaction also occurs non-enzymatically [35–37]. The mechanism has not yet been fully established. A proposal, consistent with the regiochemistry of the reaction [31, 38–40], with the observation of facile H/D exchange at the C-7 methyl group of lumazine [38, 40–42], with the facile nucleophilic addition to C-7 of lumazine [36, 43, 44], and with NMR studies [45–47], is outlined in Fig. 11 [39–40].

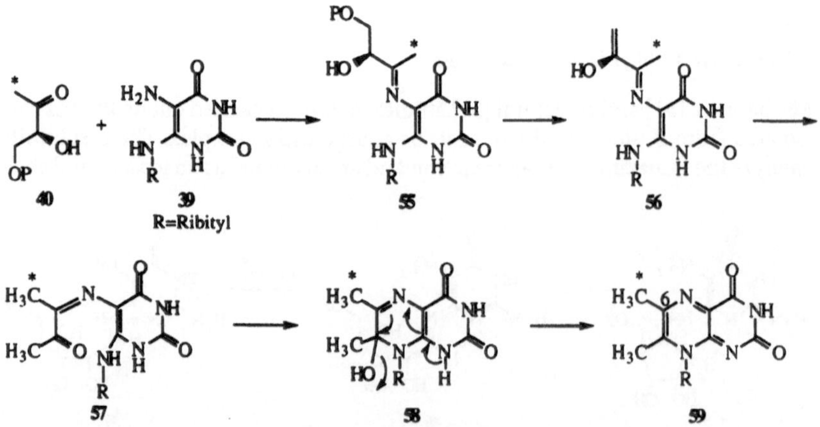

Fig. 10. Mechanistic proposal for 6,7-Dimethyl-8-ribityl lumazine synthase

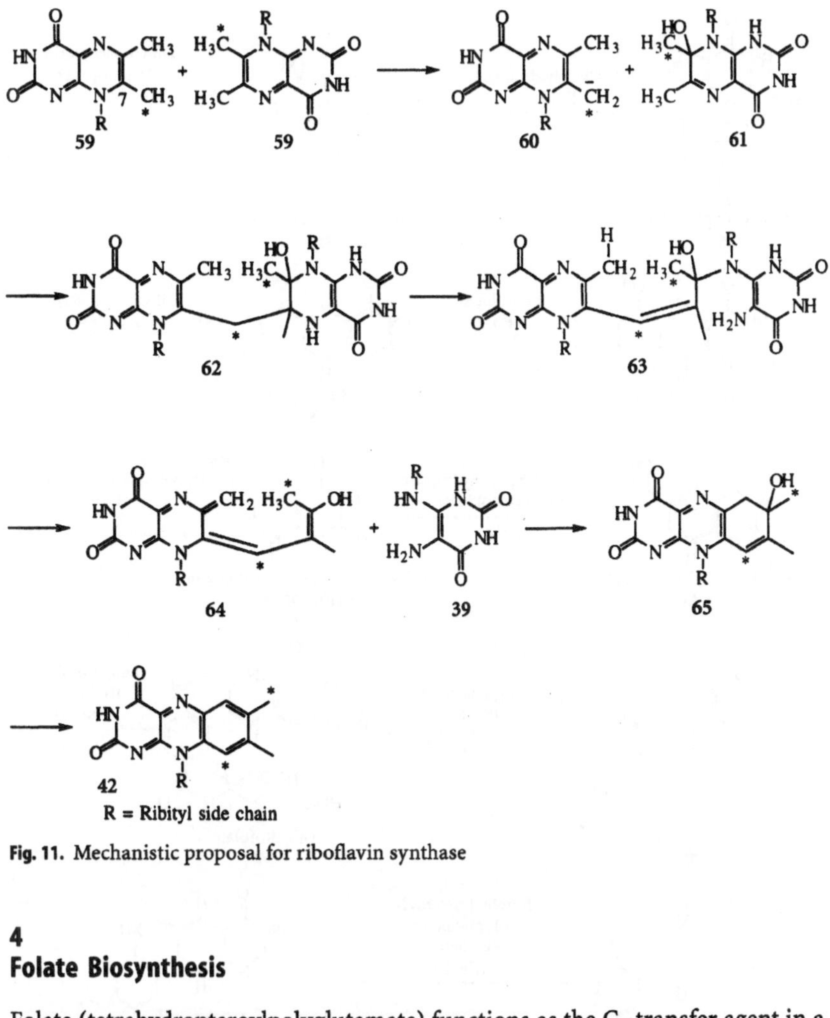

R = Ribityl side chain

Fig. 11. Mechanistic proposal for riboflavin synthase

4 Folate Biosynthesis

Folate (tetrahydropteroylpolyglutamate) functions as the C_1-transfer agent in a variety of important metabolic processes [48].

4.1 Biosynthetic Pathway

The folate biosynthetic pathway is shown in Fig. 12 [49]. Loss of C-8 from guanosine triphosphate followed by rearrangement gives **66**. Dephosphorylation, cleavage of the side chain, and pyrophosphorylation gives **70**. Displacement of the pyrophosphate by *p*-aminobenzoic acid (formed from chorismic acid by amination and loss of pyruvate- see Fig. 13) followed by condensation with glutamate and reduction gives tetrahydrofolate **73**. Polyglutamylation at the γ-carboxy group gives the biochemically active form of the cofactor.

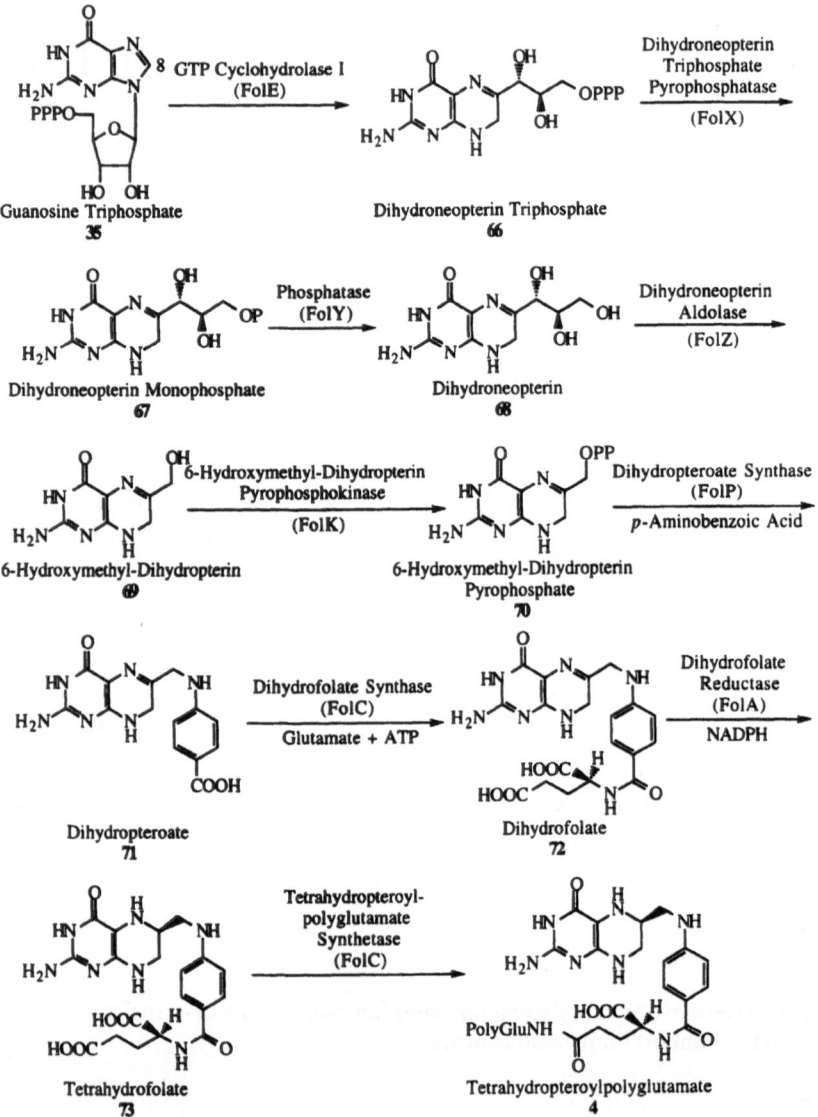

Fig. 12. The folate biosynthetic pathway

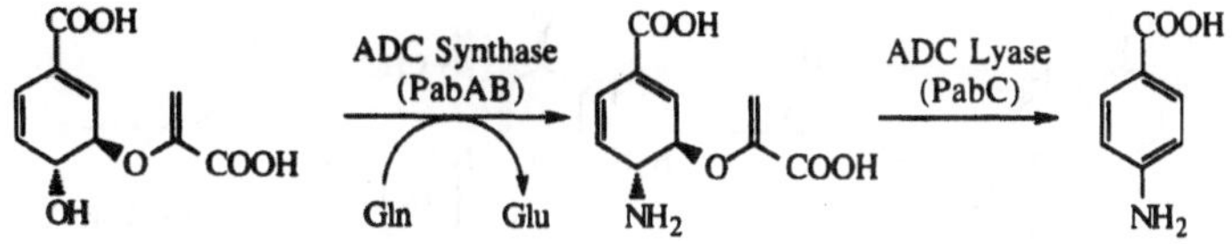

Fig. 13. The p-aminobenzoic acid biosynthetic pathway

4.2
Enzyme Overexpression and Purification

The current status of the overexpression and purification of the folate bio-synthetic enzymes is summarized in the table.

Gene	Sequenced?	Overexpressed?	Purified?	Other source [4]	Reference
folE	yes	yes	yes		50–52
folX	no	no	partial	no	53
folY	no	no	partial	no	53
folZ	no	no	partial	yes	54, 55
folK	yes	yes	yes		56
folP	yes	yes	yes		57
folC	yes	yes	yes		58, 59
folA	yes	yes	yes		60, 61
pabA	yes	yes	yes		62, 63
pabB	yes	yes	yes		62, 63
pabC	yes	yes	yes		64, 65

4.3
Mechanistic Highlights

4.3.1 GTP Cyclohydrolase I

A mechanistic proposal, supported by the crystal structure of the enzyme with bound deoxyguanosine triphosphate, is outlined in Fig. 14 [66, 67].

4.3.2
Dihydroneopterin Aldolase

A mechanistic proposal is outlined in Fig. 15. No mechanistic studies have been carried out on this enzyme.

4.3.3
Dihydropteroate Synthase

This enzyme is the target of the sulfonamide antibiotics [68]. The mechanism has not been determined. S_N1 or S_N2 mechanisms are possible.

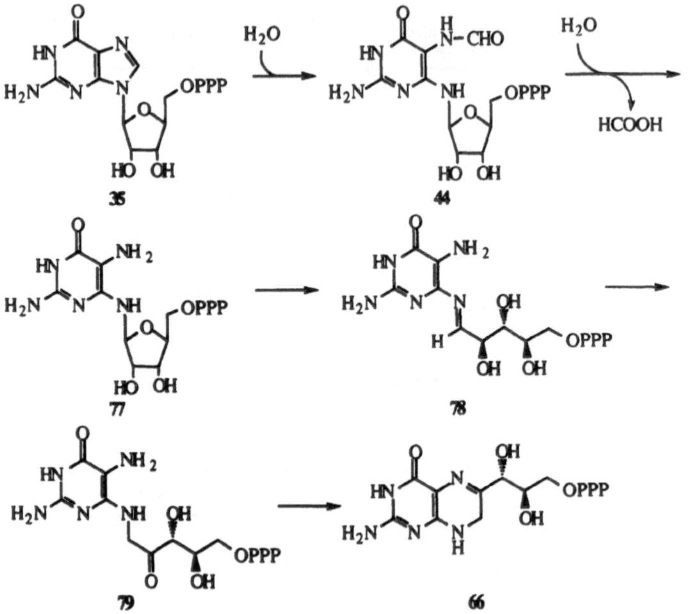

Fig. 14. Mechanistic proposal for GTP cyclohydrolase I

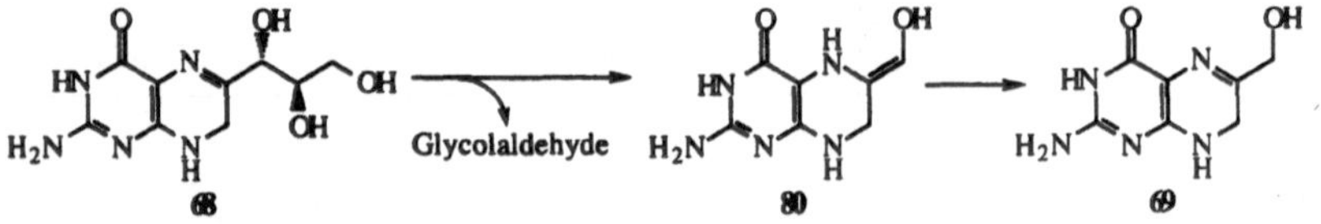

Fig. 15. Mechanistic proposal for dihydroneopterin aldolase

4.3.4
ADC Synthase

A mechanistic proposal is outlined in Fig. 16. It is not known whether the carbinolamine 82 or free ammonia reacts with chorismate [69], or whether the addition of the amine involves the initial addition of an active site nucleophile to chorismate [70–74].

4.3.5
ADC Lyase

A mechanistic proposal for this pyridoxal-dependent enzyme is outlined in Fig. 17 [64]. No mechanistic studies have been reported.

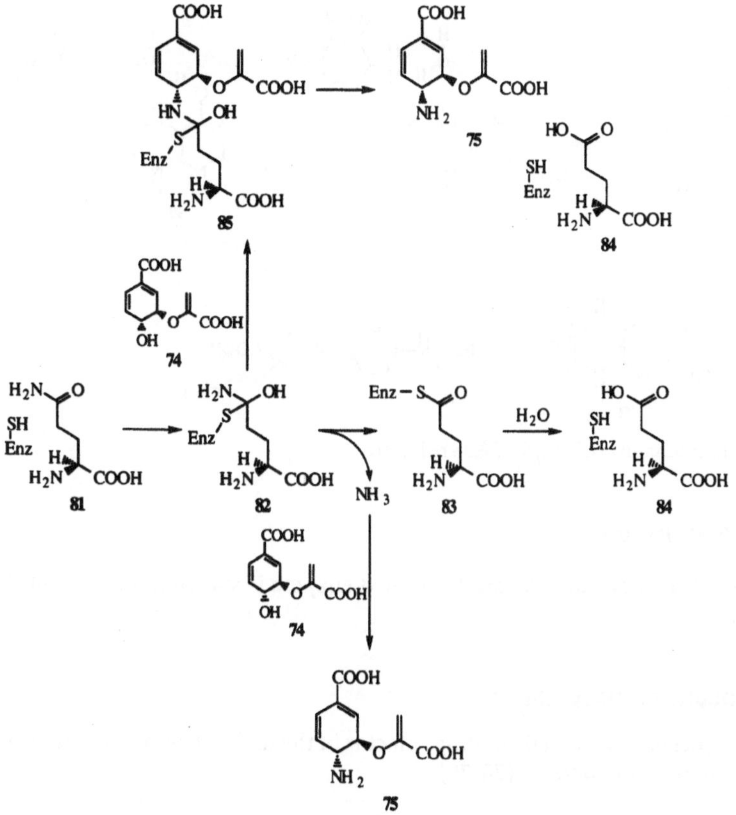

Fig. 16. Mechanistic proposal for ADC synthase

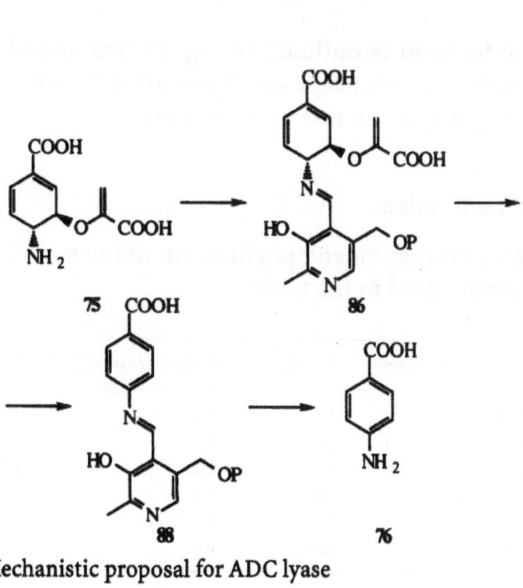

Fig. 17. Mechanistic proposal for ADC lyase

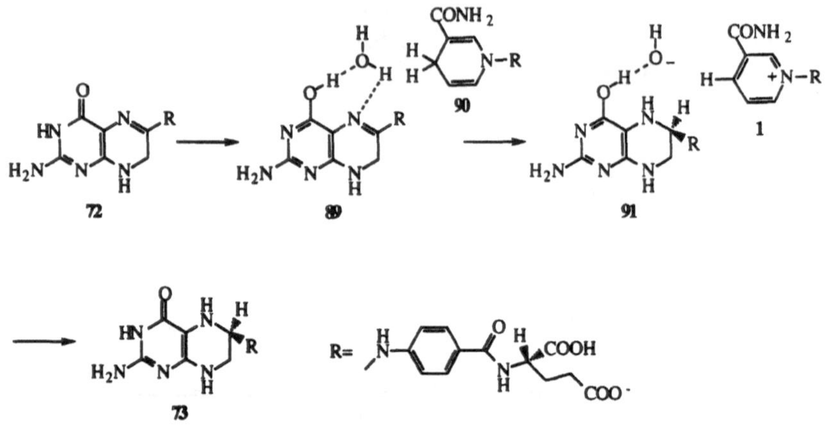

Fig. 18. The mechanism of dihydrofolate reductase

4.3.6
Dihydrofolate Reductase

The mechanism of this intensively-studied enzyme is shown in Fig. 18 [48, 75, 76, 77].

5
Molybdopterin Biosynthesis

Molybdopterin is the metal-binding cofactor in the molybdenum- and tungsten dependent oxidoreductases [78, 79].

5.1
Biosynthetic Pathway

The biosynthesis of molybdopterin is outlined in Fig. 19. The initial step involves rearrangement of guanosine triphosphate to precursor Z. Sulfur transfer followed by metallation and guanylation gives the cofactor.

5.2
Enzyme Overexpression and Purification

The current status of the overexpression and purification of the molybdopterin biosynthetic enzymes is summarized in the table.

Gene	Sequenced?	Overexpressed?	Purified?	Other source? [4]	Reference
moaABC	yes	no	no	no	80
moaDE	yes	no	yes	no	80, 81
moeB	yes	yes	yes	no	80, 82
mobAB	yes	no	no	no	80
modABCD	yes	no	no	no	80
mogA	yes	no	no	no	80, 83

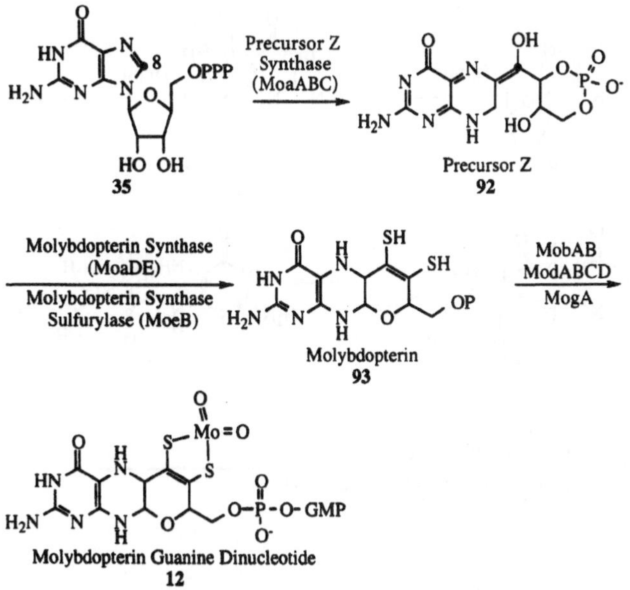

Fig. 19. The molybdopterin biosynthetic pathway

5.3
Mechanistic Highlights

5.3.1
Precursor Z Synthase

Mechanistic studies on the formation of the molybdopterin cofactor are still at an early stage. Conversion of guanosine, presumably as the triphosphate, to precursor Z occurs with retention of C-8 [84]. A possible mechanism for this third type of cyclohydrolase, which is consistent with the labeling experiment, is outlined in Fig. 20. (The other two types of cyclohydrolase are cyclohydrolase I for folate biosynthesis and cyclohydrolase II for riboflavin biosynthesis. In both cases, C8 is removed as formate.)

5.3.2
Molybdopterin Synthase

The mechanism for the conversion of precursor Z to molybdopterin is not known. Sulfur, from an undetermined source, is transferred from MoaD to precursor Z [81, 85].

A mechanistic proposal for the sulfur transfer, tentatively based on the sequence similarity between MoeB and the ubiquitin activating enzyme is outlined in Fig. 21 [80]. MoeB has high sequence similarity to ThiF [86], suggesting that similar sulfur transfer chemistry may occur during the biosynthesis of the thiazole moiety of thiamin (see section 6 of this review).

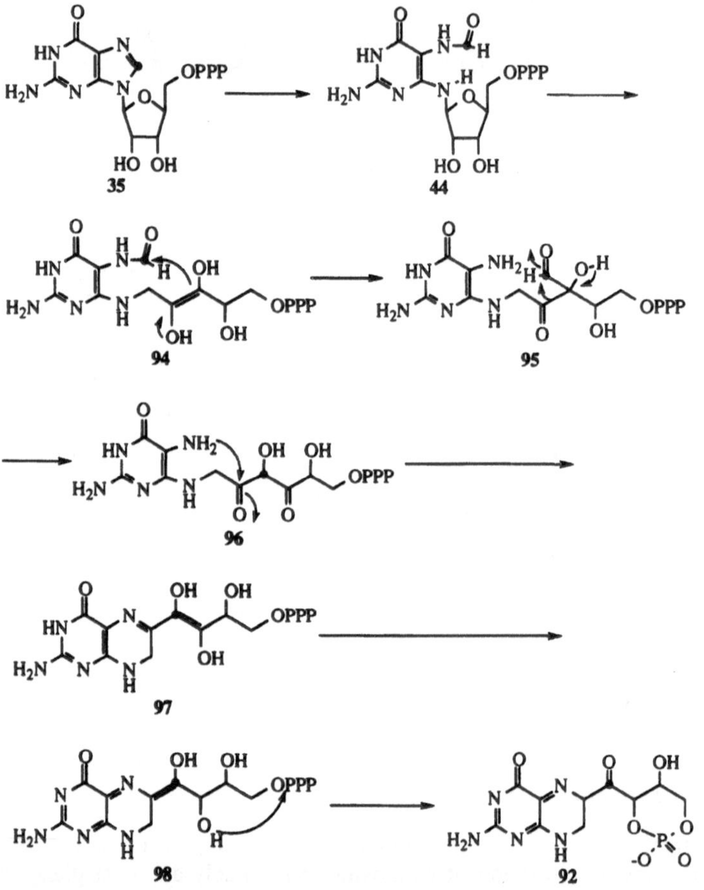

Fig. 20. Mechanistic proposal for precursor Z synthase

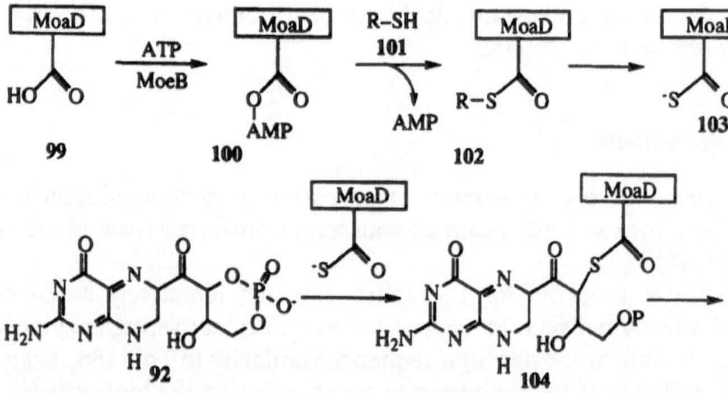

Fig. 21. Mechanistic proposal for molybdopterin synthase

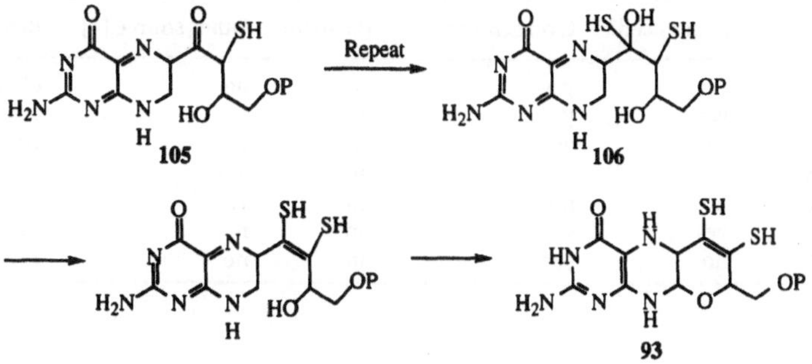

Fig. 21 (continued)

6
Thiamin Biosynthesis

Thiamin is involved in the stabilization of the acyl carbanion intermediate and is particularly important in carbohydrate metabolism [87].

6.1
Biosynthetic Pathway

The thiamin biosynthetic pathway is outlined in Fig. 22 [88]. Overall the pathway involves the separate synthesis of the thiazole 111 and the pyrimidine 114 which are then coupled. 1-Deoxy-D-xylulose-5-phosphate (see sections 11 and 12 of this review) is the precursor to the five carbon unit of the thiazole [89], cysteine is the sulfur source [90, 91], and the C2-N3 atoms of the thiazole are derived from the α-carbon and the amino group of tyrosine [92–94]. The pyrimidine is derived from 5-aminoimidazole riboside (AIRs), an intermediate on the purine biosynthetic pathway. This reaction involves a complex rearrangement in which the C4' carbon of AIRs is inserted into the C4–C5 imidazole double bond, converting the imidazole to a pyrimidine, and the C2' carbon of AIRs is used to methylate the C2 position of the imidazole [95–98].

6.2
Enzyme Overexpression and Purification

The current status of the overexpression and purification of the thiamin biosynthetic enzymes is summarized in the table.

Gene	Sequenced?	Overexpressed?	Purified?	Other source [4]	Reference
thiC	yes	yes	yes		86, 99, 100
thiE	yes	yes	yes		86, 101, 102
thiF	yes	yes	yes		86, 103
thiG$_1$	yes	yes	yes		86, 103
thiG$_2$	yes	yes	yes		86, 103

Gene	Sequenced?	Overexpressed?	Purified?	Other source [4]	Reference
thiH	yes	yes	no	no	86, 103
thiI	yes	yes	yes		104
thiJ	yes	yes	no	no	105, 106
thiM	no	no	no	yes	107, 108
thiD	no	no	no	yes	108–110
thiL	yes	yes	no	no	111–113
thiK	no	yes	no	no	111, 112

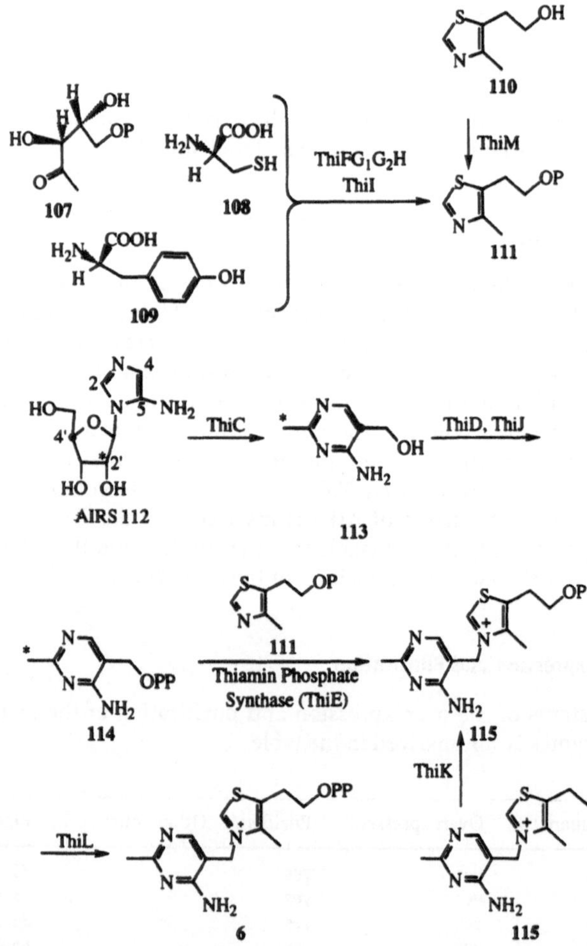

Fig. 22. The thiamin biosynthetic pathway

6.3
Mechanistic Highlights

6.3.1
Thiazole Formation

Cell free biosynthesis of the thiazole has not yet been established. A mechanistic proposal based on the sequence similarity between ThiF, MoeB, and the ubiquitin activating enzyme and on the occurrence of a putative zinc ion-binding motif in ThiF [80] is outlined in Fig. 23 [114].

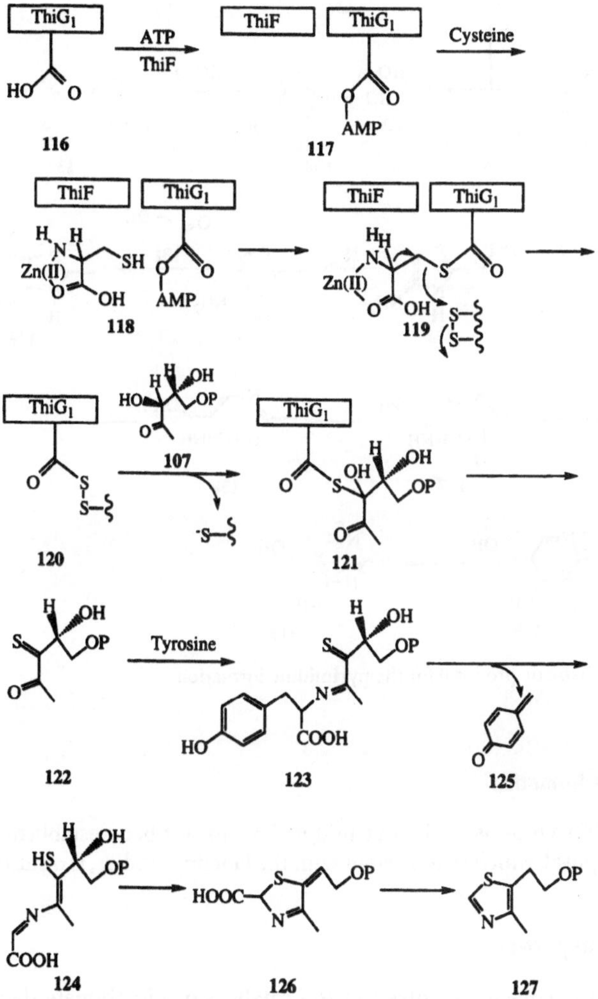

Fig. 23. Mechanistic proposal for the thiazole formation

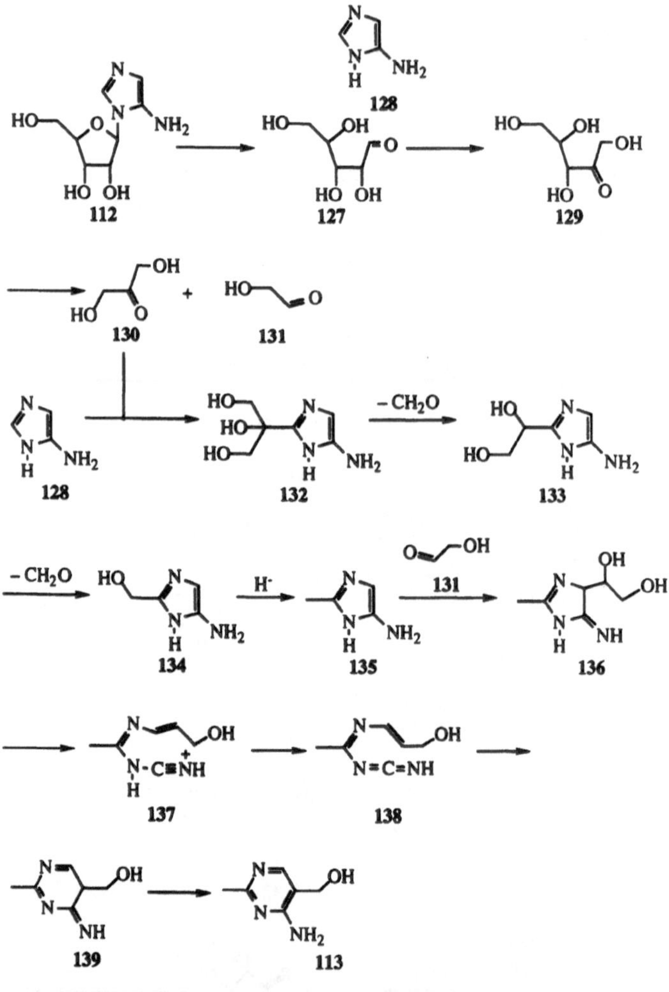

Fig. 24. Mechanistic proposal for the pyrimidine formation

6.3.2
Pyrimidine Formation

Cell free biosynthesis of the pyrimidine has not yet been established. A mechanistic proposal, which is consistent with the labeling studies, is outlined in Fig. 24.

7
Biotin Biosynthesis

Biotin is the cofactor involved in the catalysis of bicarbonate-dependent carboxylation reactions.

7.1
Biosynthetic Pathway

The biotin biosynthetic pathway is outlined in Fig. 25 [115]. Pimeloyl-CoA, synthesized by a variation of the fatty acid biosynthesis pathway, is condensed with alanine to give 142.

A transamination reaction, using S-adenosylmethionine as the amine donor, gives 143. Urea formation, followed by sulfur insertion into the C-1H and the C-4H bonds gives biotin.

7.2
Enzyme Overexpression and Purification

The current status of the overexpression and purification of the biotin biosynthetic enzymes is summarized in the table [115, 116].

Gene	Sequenced?	Overexpressed?	Purified?	Other source [4]	Reference
bioC	yes	no	no	no	117
bioH	yes	no	no	no	117
bioF	yes	yes	yes	yes	118, 119
bioA	yes	no	yes		120
bioD	yes	yes	yes		121
bioB	yes	yes	yes		122

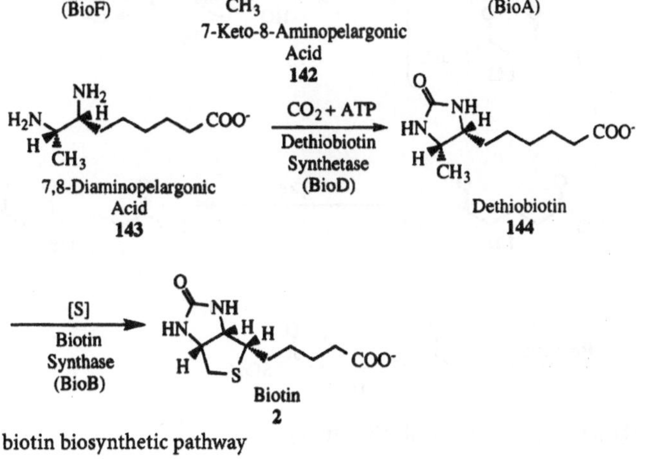

Fig. 25. The biotin biosynthetic pathway

7.3
Mechanistic Highlights

7.3.1
Pimeloyl CoA Synthase

Labeling studies suggest that pimeloyl CoA is synthesized by a modified fatty acid pathway (Fig. 26) [117, 123, 124]. Cell free enzymatic activity has not been reported. Pimeloyl CoA is also synthesized from pimelic acid in *Bacillus sphaericus* [125].

7.3.2
KAPA Synthase

The mechanistic proposal for this pyridoxal phosphate-dependent enzyme is outlined in Fig. 27 and is supported by isotope exchange experiments [118].

7.3.3
DAPA Synthase

This PLP-dependent enzyme is unique in that it requires SAM rather than glutamic acid as the amino group donor.

Although mechanistic studies have not been carried out, the reaction is likely to follow the standard transamination mechanism (Fig. 28) [126, 127].

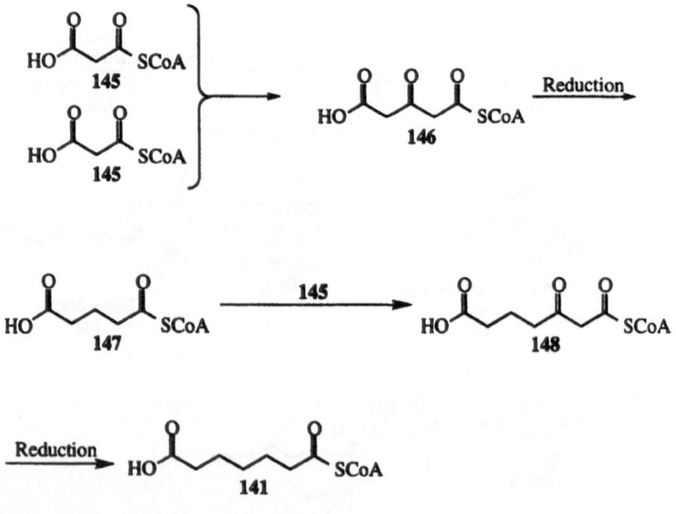

Fig. 26. The pimeloyl-CoA biosynthetic pathway

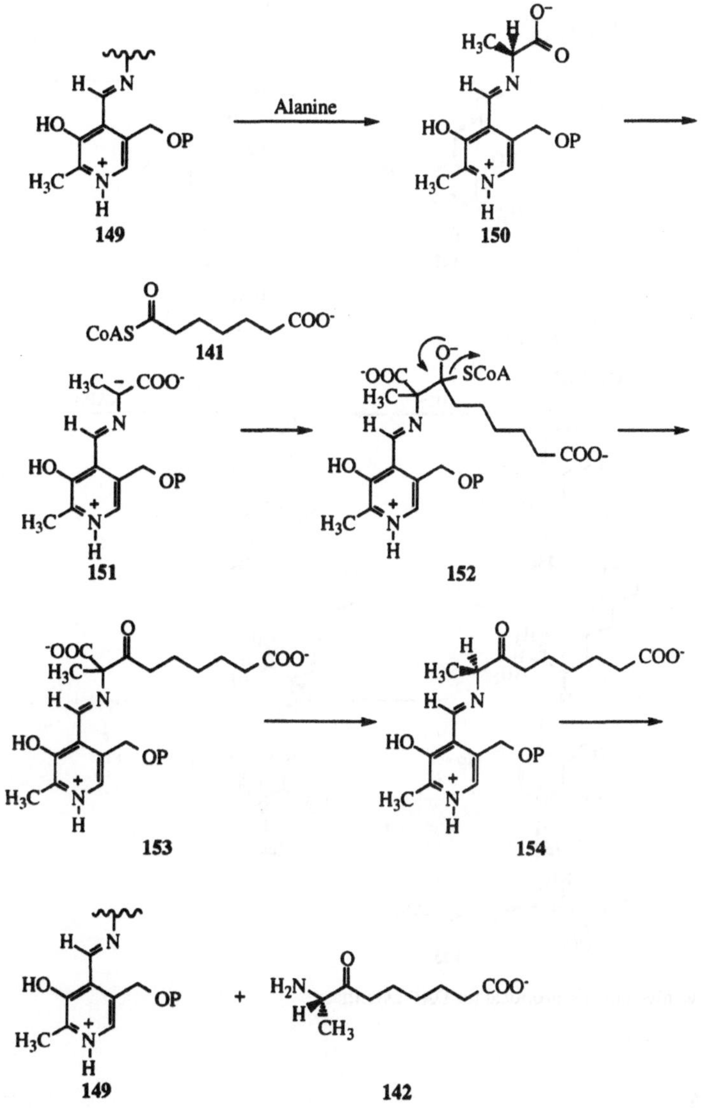

Fig. 27. Mechanistic proposal for KAPA synthase

7.3.4
Dethiobiotin Synthetase

The mechanistic proposal for this reaction is outlined in Fig. 29 and is supported by the demonstration that 162 is an intermediate [128] and by the crystal structure [129, 130].

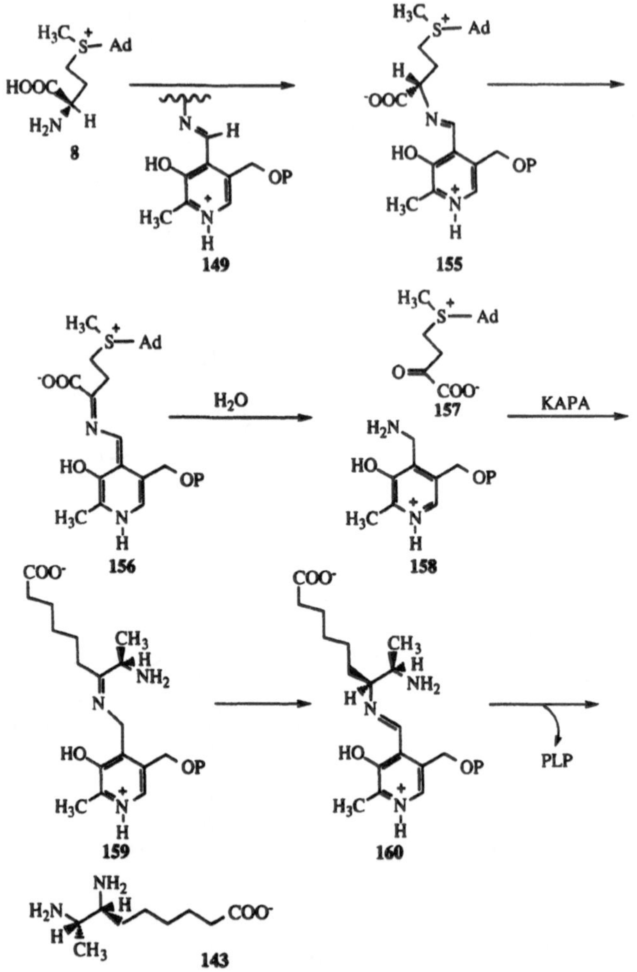

Fig. 28. Mechanistic proposal for DAPA synthase

7.3.5
Biotin Synthase

This very interesting reaction involves the formal insertion of a sulfur atom into two unactivated CH bonds. The purified protein contains a [2Fe–2S] cluster [131]. A defined system, consisting of biotin synthase, flavodoxin, flavodoxin reductase, fructose 1,6-bisphosphate, cysteine, DTT, NADPH, ferrous ion, and SAM, capable of catalyzing the conversion of dethiobiotin to biotin has been characterized [132–134]. This system is still incomplete and gives a maximum of two moles of biotin per mole of biotin synthase.

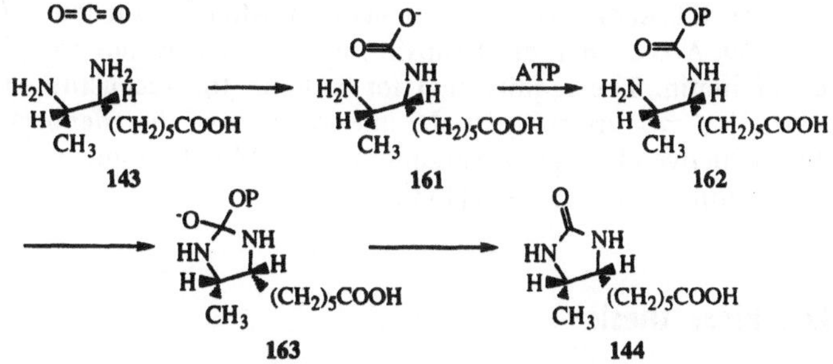

Fig. 29. Mechanistic proposal for dethiobiotin synthetase

The sulfur insertion occurs with scrambling of stereochemistry at C-1 and with retention at C-4 [135,136]. The protons at C-2 and at C-3 are not lost during the reaction. 1-Mercaptodethiobiotin may be an intermediate [137, 138]. In the reconstituted system, sulfur from cysteine or SAM was not incorporated into biotin [132] suggesting that the iron sulfur cluster is the sulfur source.

A mechanistic hypothesis is outlined in Fig. 30. Hydrogen atom abstraction from C-1 by an adenosyl radical (or by a protein radical formed from the adeno-

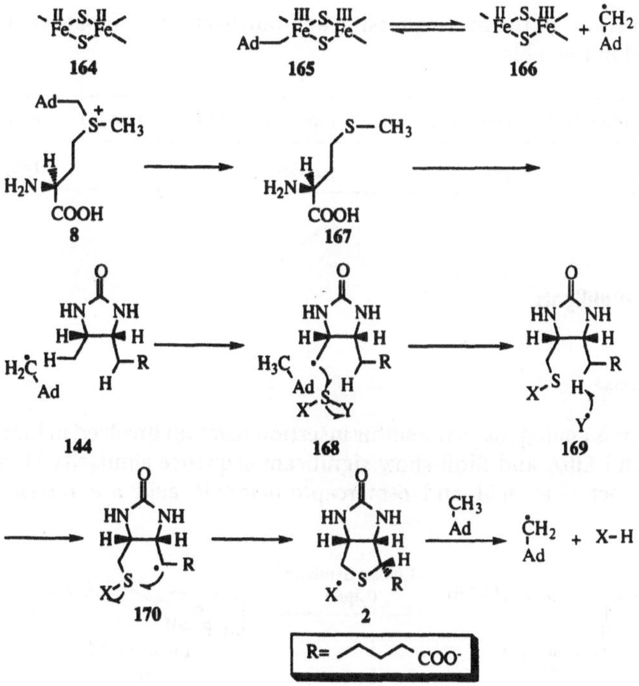

Fig. 30. Mechanistic proposal for biotin synthase

syl radical) gives **168**. Radical substitution on the undefined sulfur source (X-S-Y) would give **169**. A second radical substitution reaction would complete the formation of biotin. The requirement for SAM in the reconstituted biotin synthase reaction mixture suggests the intermediacy of the adenosyl radical [132, 139]. Reductive cleavage of sulfonium salts [140–142] and radical substitutions at sulfur are precedented [143].

8 Lipoic Acid Biosynthesis

Lipoic acid is a cofactor involved in acyl group transfer and 2-electron redox reactions.

8.1 Biosynthetic Pathway

The biosynthesis of lipoic acid from octanoic acid is outlined in Fig. 31 [115] and involves the insertion of sulfur into two unactivated CH bonds [144].

8.2 Enzyme Overexpression and Purification

The current status of the overexpression and purification of lipoate synthase is summarized in the table.

Gene	Sequenced?	Overexpressed?	Purified?	Other source [4]	Reference
lipA	yes	yes	partial	no	145, 146

8.3 Mechanistic Highlights

8.3.1 *Lipoate Synthase*

This reaction is analogous to the sulfur insertion reaction involved in biotin biosynthesis, and LipA and BioB show significant sequence similarity [145, 146]. 8-Mercapto-octanoic acid and 6-mercapto-octanoic acid are intermediates

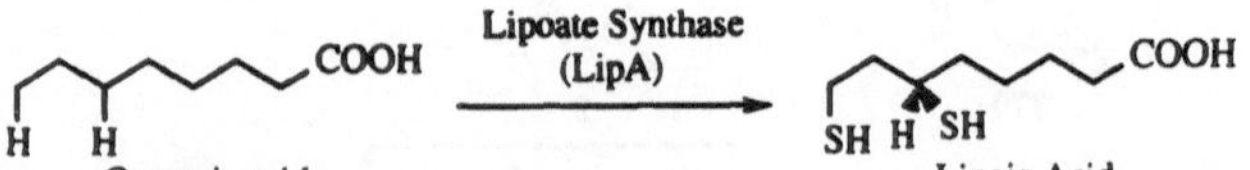

Octanoic acid **171** Lipoic Acid **11**

Fig. 31. The lipoic acid biosynthetic pathway

[147, 148]. The sulfur insertion into the C-6H bond occurs with inversion of stereochemistry [149]. Cysteine is the sulfur source [150]. Lipoic acid synthesis in a defined cell free system has not yet been accomplished.

9
Pantothenic Acid and Coenzyme A Biosynthesis

Coenzyme A and pantothenic acid function as acyl group carriers and play a key role in biochemical Claisen condensation reactions.

9.1
Biosynthetic Pathway

The biosynthesis of panthothenate and coenzyme A is outlined in Fig. 32 [151]. Hydroxymethylation of 172 followed by reduction and condensation with β-alanine (formed by the decarboxylation of aspartate) gives pantothenic acid. Phosphorylation of panthothenate, followed by condensation with cysteine and decarboxylation gives 179. Adenylation, followed by a final phosphorylation at the ribose 3'-hydroxyl, completes the biosynthesis.

9.2
Enzyme Overexpression and Purification

The current status of the overexpression and purification of the panthothenate and coenzyme A biosynthetic enzymes is summarized in the table.

Gene	Sequenced?	Overexpressed?	Purified?	Other source [4]	Reference
panB	yes	yes	yes		152
panE	no	no	partial	yes	153, 154
panD	no	no	yes		155
panC	no	no	yes		156
coaA	yes	yes	yes		157
panW	no	no	no	yes	158
panX	no	no	yes		159
panY	no	no	no	yes	160, 161
panZ	no	no	no	yes	161

9.3
Mechanistic Highlights

9.3.1
Ketopantoate Hydroxymethyltransferase

The purified enzyme requires a methylene tetrahydrofolate cofactor [162]. The hydroxymethylation reaction proceeds with inversion [163] (Fig. 33) and probably occurs by a mechanism analogous to serine hydroxymethyl transferase [48].

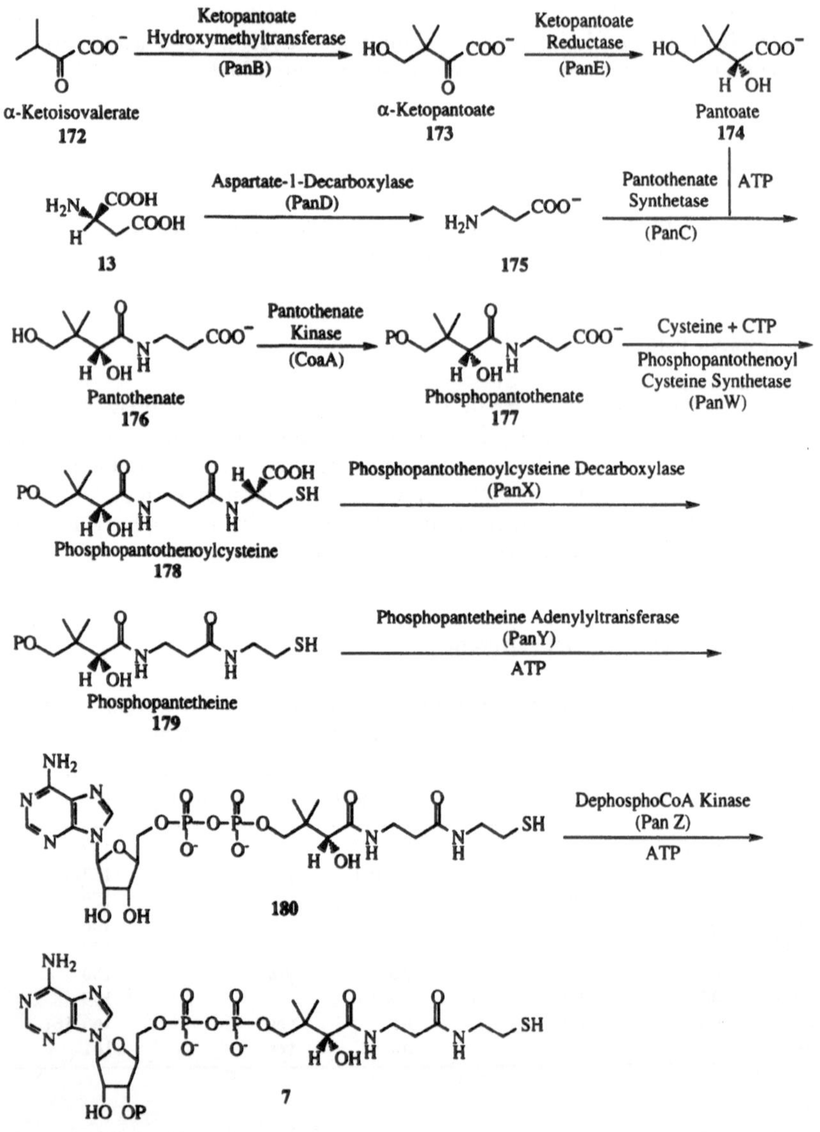

Fig. 32. The pantothenic acid and coenzyme A biosynthetic pathway

9.3.2
Aspartate Decarboxylase

This enzyme contains a pyruvamide cofactor. No mechanistic studies have been reported. A proposal, analogous to the mechanism of histidine decarboxylase, is outlined in Fig. 34 [164].

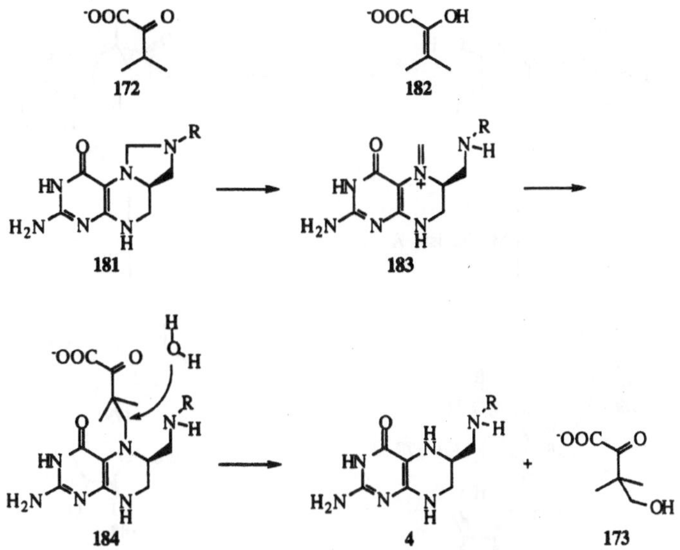

Fig. 33. Mechanistic proposal for ketopantoate hydroxymethyltransferase

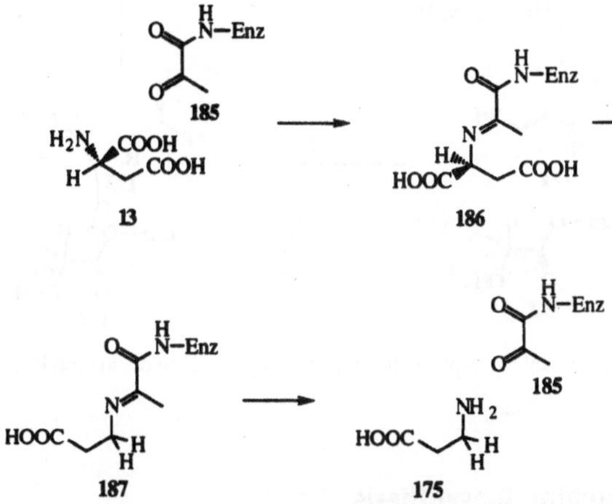

Fig. 34. Mechanistic proposal for aspartate decarboxylase

9.3.3
4-Phosphopantothenoylcysteine Decarboxylase

This interesting enzyme contains a pyruvoyl cofactor [159, 164, 165]. The mechanism is unknown. Decarboxylation proceeds with retention of stereochemistry [166]. Two proposals are outlined in Fig. 35.

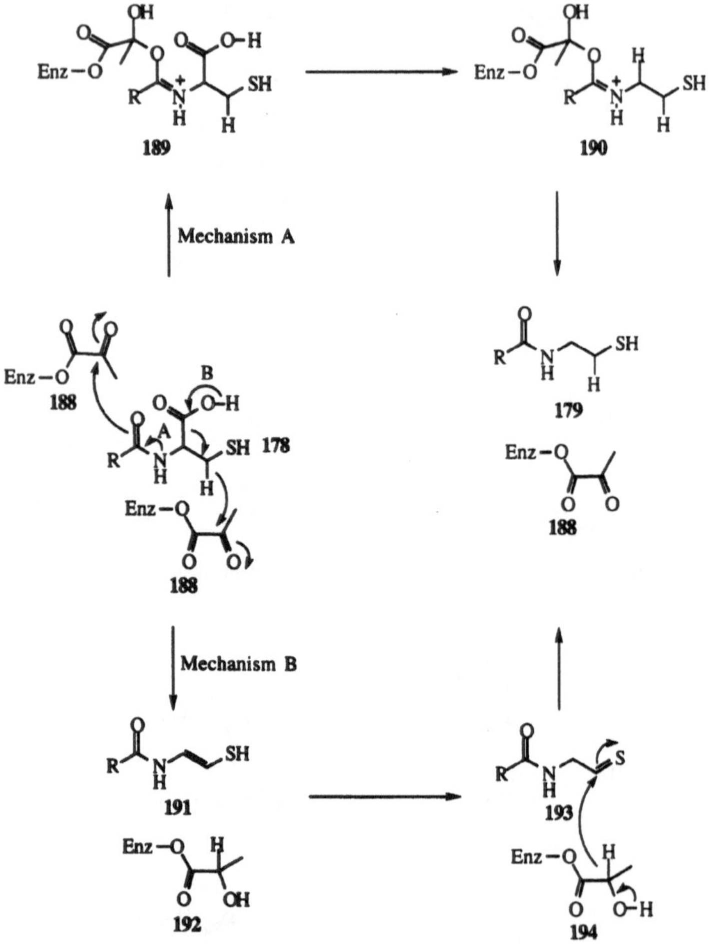

Fig. 35. Mechanistic proposal for 4-phosphopantothenoylcysteine decarboxylase

10
S-Adenosylmethionine Biosynthesis

S-adenosylmethionine (SAM) is the cofactor involved in most biochemical methylation reactions.

10.1
Biosynthetic Pathway

The biosynthesis of SAM involves the alkylation of methionine by ATP and is outlined in Fig. 36.

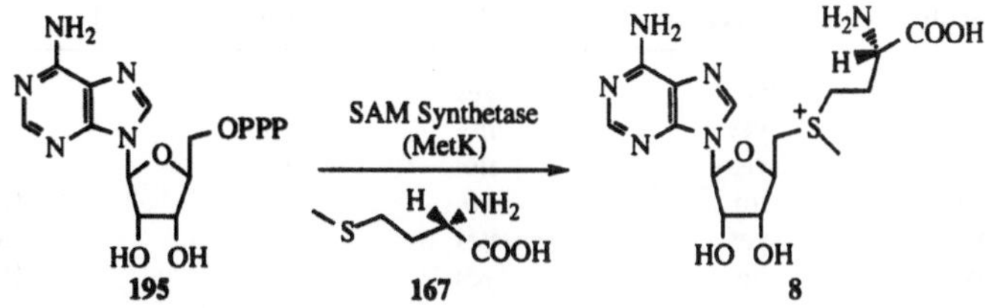

Fig. 36. The S-adenosylmethionine biosynthetic pathway

10.2
Enzyme Overexpression and Purfication

The current status of the overexpression and purification of SAM synthetase is summarized in the table.

Gene	Sequenced?	Overexpressed?	Purified?	Other source [4]	Reference
metK	yes	yes	yes		167

10.3
Mechanistic Highlights

10.3.1
SAM Synthetase

Stereochemical studies [168] and isotope effects [169] are consistent with an S_N2 mechanism. The crystal structure has been determined [170].

11
Pyridoxal Phosphate Biosynthesis

Pyridoxal phosphate is the cofactor involved in the stabilization of carbanions at the α and β positions of amino acids [126].

11.1
Biosynthetic Pathway

The pyridoxal biosynthetic pathway is outlined in Fig. 37. Oxidation of **196** followed by transamination gives 4-hydroxy-L-threonine-4-phosphate **199** [171]. Condensation with 1-deoxy-D-xylulose (see also ubiquinone and thiamin sections) and a final oxidation gives the cofactor [172–175].

11.2
Enzyme Overexpression and Purification

The current status of the overexpression and purification of the pyridoxal phosphate biosynthetic enzymes is summarized in the table [176].

Gene	Sequenced?	Overexpressed?	Purified?	Other source [4]	Reference
epd	yes	yes	yes		177
pdxB	yes	yes	no	no	178
serC	yes	yes	no	no	178
pdxAJ	yes	no	no	no	176, 179
pdxH	yes	yes	yes		180
pdxK	yes	no	yes		181

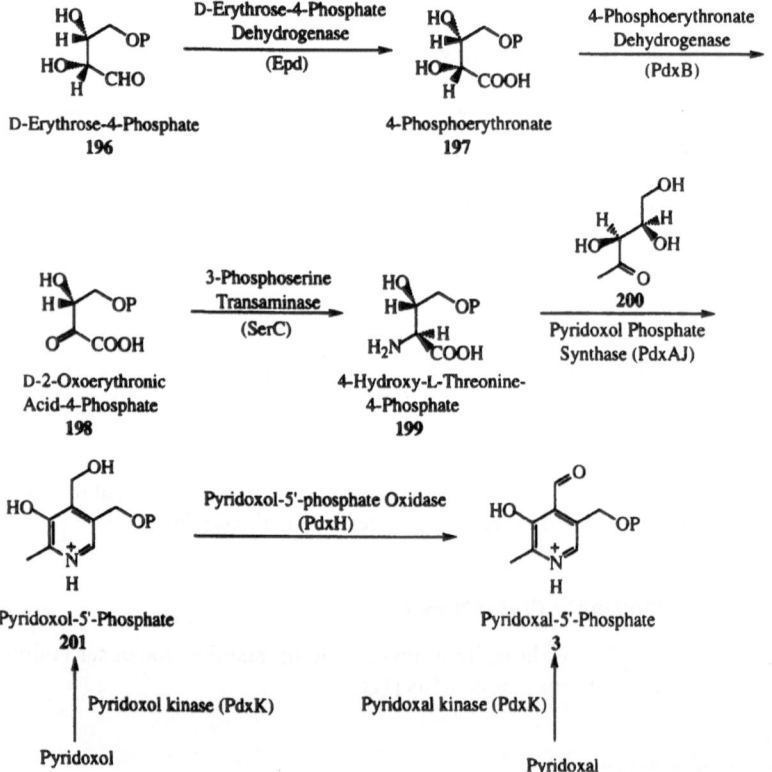

Fig. 37. The pyridoxal phosphate biosynthetic pathway

11.3
Mechanistic Highlights

11.3.1
Pyridoxol phosphate synthase

The pyridoxol phosphate synthase activity has not yet been reconstituted in a cell free system. A mechanistic proposal is outlined in Fig. 38 [176].

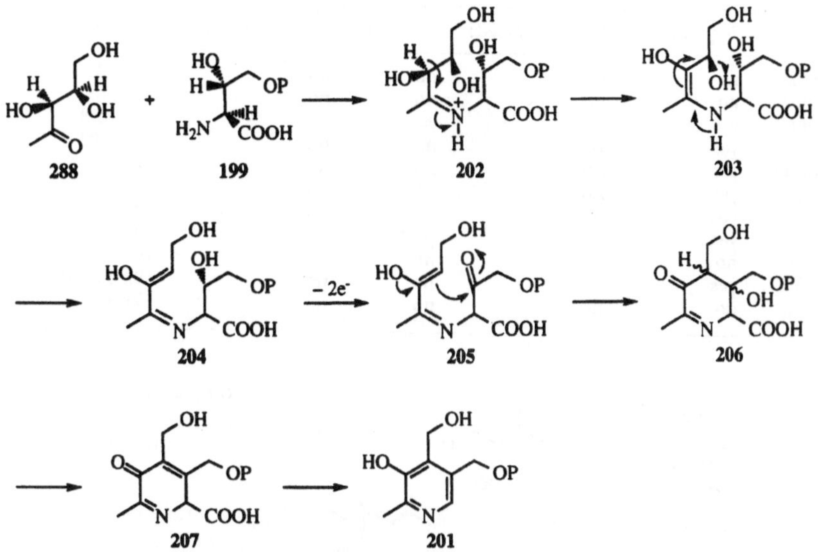

Fig. 38. Mechanistic proposal for pyridoxal phosphate synthase

12
Ubiquinone Biosynthesis

Ubiquinone mediates electron transfer between dehydrogenases and the cyto-chromes.

12.1
Biosynthetic Pathway

The biosynthesis of ubiquinone is outlined in Fig. 39 [182]. Elimination of pyruvate from chorismate followed by prenylation and decarboxylation gives 2-octaprenyl phenol. A sequence of three hydroxylation/methylation reactions completes the biosynthesis.

The isopentenyl pyrophosphate and the dimethylallyl pyrophosphate precursors to the octaprenyl moiety are derived from 1-deoxy-D-xylulose-5-phosphate rather than from mevalonic acid (see also thiamin and pyridoxal sections) [183, 184].

12.2
Enzyme Overexpression and Purification

The current status of the overexpression and purification of the ubiquinone biosynthetic enzymes is summarized in the table.

T. P. Begley et al.

Gene	Sequenced?	Overexpressed?	Purified?	Other source [4]	Reference
ubiJ	no	no	no	no	183, 185
ubiK	no	no	no	no	183, 185
ubiC	yes	yes	yes		186–188
ubiA	yes	yes	no	no	189–192
ubiD	yes	no	no	no	193
ubiX	yes	no	no	no	182
ubiB	yes	no	no	no	194
ubiG	yes	no	partial	no	194–196
ubiH	no	no	no	no	194, 182
ubiE	yes	no	no	no	182, 194, 197
ubiF	no	no	no	no	194, 182

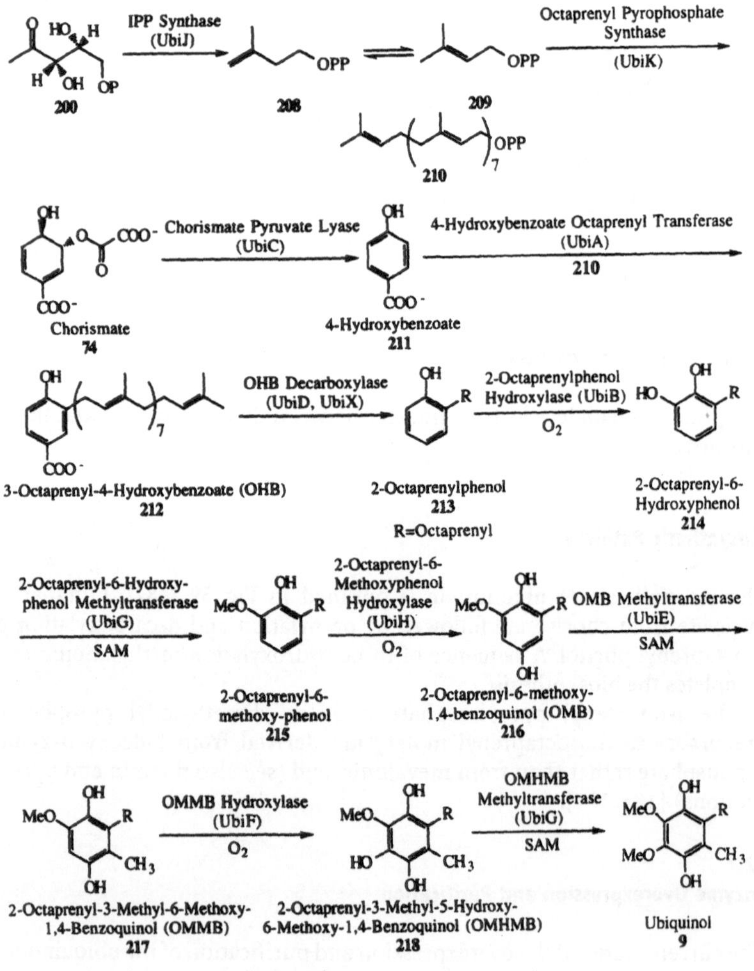

IPP Synthase (UbiJ) — 200 → 208 ⇌ 209 — Octaprenyl Pyrophosphate Synthase (UbiK) → 210

Chorismate Pyruvate Lyase (UbiC) — Chorismate 74 → 4-Hydroxybenzoate 211 — 4-Hydroxybenzoate Octaprenyl Transferase (UbiA) 210 →

OHB Decarboxylase (UbiD, UbiX) — 3-Octaprenyl-4-Hydroxybenzoate (OHB) 212 → 2-Octaprenylphenol 213 (R=Octaprenyl) — 2-Octaprenylphenol Hydroxylase (UbiB) O_2 — 2-Octaprenyl-6-Hydroxyphenol 214

2-Octaprenyl-6-Hydroxy-phenol Methyltransferase (UbiG) SAM → 2-Octaprenyl-6-methoxy-phenol 215 — 2-Octaprenyl-6-Methoxyphenol Hydroxylase (UbiH) O_2 — 2-Octaprenyl-6-methoxy-1,4-benzoquinol (OMB) 216 — OMB Methyltransferase (UbiE) SAM →

OMMB Hydroxylase (UbiF) O_2 — 2-Octaprenyl-3-Methyl-6-Methoxy-1,4-Benzoquinol (OMMB) 217 → 2-Octaprenyl-3-Methyl-5-Hydroxy-6-Methoxy-1,4-Benzoquinol (OMHMB) 218 — OMHMB Methyltransferase (UbiG) SAM → Ubiquinol 9

Fig. 39. The ubiquinone biosynthetic pathway

12.3
Mechanistic Highlights

12.3.1
4-Hydroxybenzoate Octaprenyl Transferase

The mechanism for this reaction has not been studied. A simple electrophilic substitution as outlined in Fig. 40 is likely. The reaction may involve an S_N2 attack by the phenol on the pyrophosphate or proceed via a carbocation.

12.3.2
OHB Decarboxylase

The mechanism of this reaction has not been studied. A proposal is outlined in Fig. 41.

12.3.3
2-Octaprenylphenol Monooxygenase

There are three hydroxylation reactions on the ubiquinone biosynthetic pathway. All three oxygens are derived from molecular oxygen [198] and heme is not required [199]. This suggests that these hydroxylations proceed by a mechanism analogous to that of the flavin dependent tyrosine hydroxylase (Fig. 42) [14].

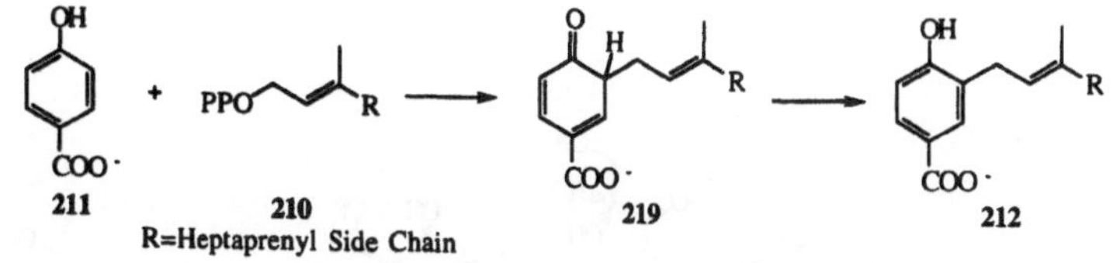

Fig. 40. Mechanistic proposal for 4-hydroxybenzoate octaprenyl transferase

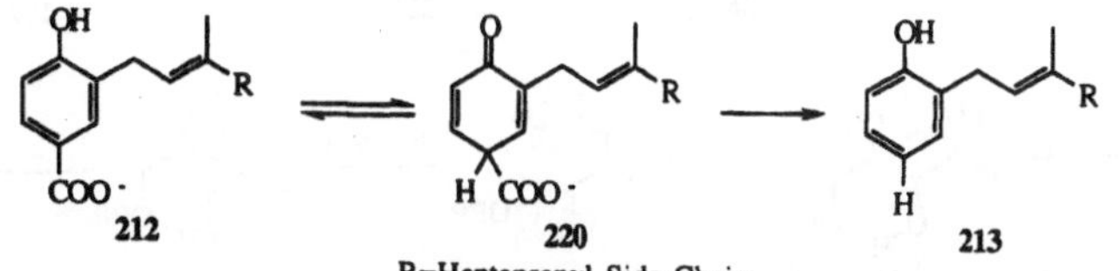

Fig. 41. Mechanistic proposal for OHB decarboxylase

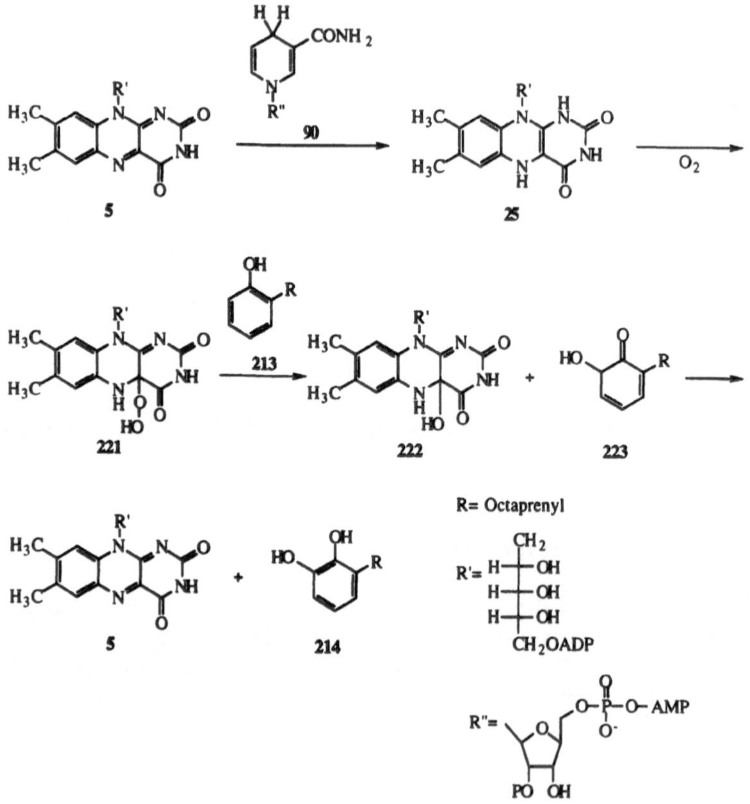

Fig. 42. Mechanistic proposal for 2-octaprenylphenol monooxygenase

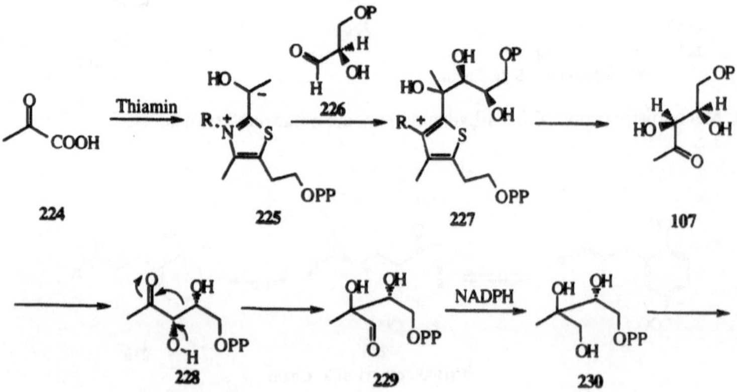

Fig. 43. Mechanistic proposal for the formation of isopentenyl pyrophosphate

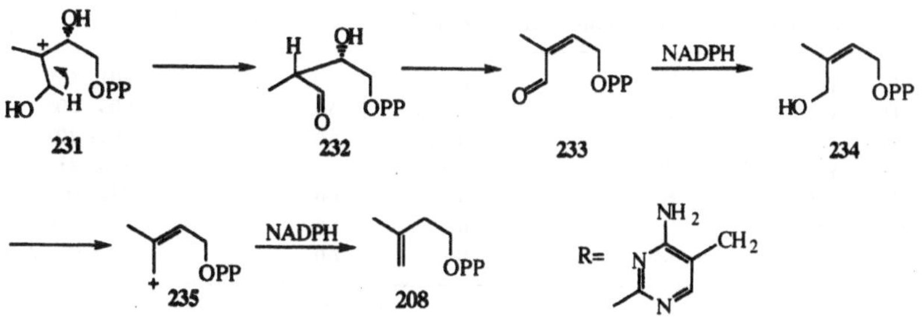

Fig. 43 (continued)

12.3.4
Formation of Isopentenyl Pyrophosphate

A proposal for the formation of 1-deoxy-D-xylulose-5-phosphate and its conversion to isopentenyl pyrophosphate is outlined in Fig. 43 [183, 185]. None of the enzymes involved have yet been cloned or characterized.

13
Menaquinone Biosynthesis

Menaquinone (vitamin K_2) mediates electron transfer between dehydrogenases and the cytochromes and is the major electron carrier cofactor during anaerobic growth of *E. coli*. Vitamin K is also an essential cofactor in the post-translational carboxylation of glutamic acid residues in several protein systems [200].

13.1
Biosynthetic Pathway

The biosynthesis of menaquinone is outlined in Fig. 44. Isomerization of chorismate to isochorismate followed by condensation with α-ketoglutarate and aromatization gives *o*-succinylbenzoic acid. Conversion of **238** to the CoA thioester, followed by cyclization, prenylation and methylation completes the biosynthesis. The biosynthesis of the prenyl side chain follows the alternative terpene biosynthetic pathway described for ubiquinone.

13.2
Enzyme Overexpression and Purification

The current status of the overexpression and purification of the menaquinone biosynthetic enzymes is summarized in the table [182].

Gene	Sequenced?	Overexpressed?	Purified?	Other source [4]	Reference
menF	yes	no	no	no	201, 202
menD	yes	yes	partial	no	203–206
menC	yes	no	partial	no	205–207
menE	yes	yes	yes		208–212
menB	yes	no	no	no	212, 213
menA	yes	no	no	no	214, 215
menG	yes	no	no	no	197, 212, 216

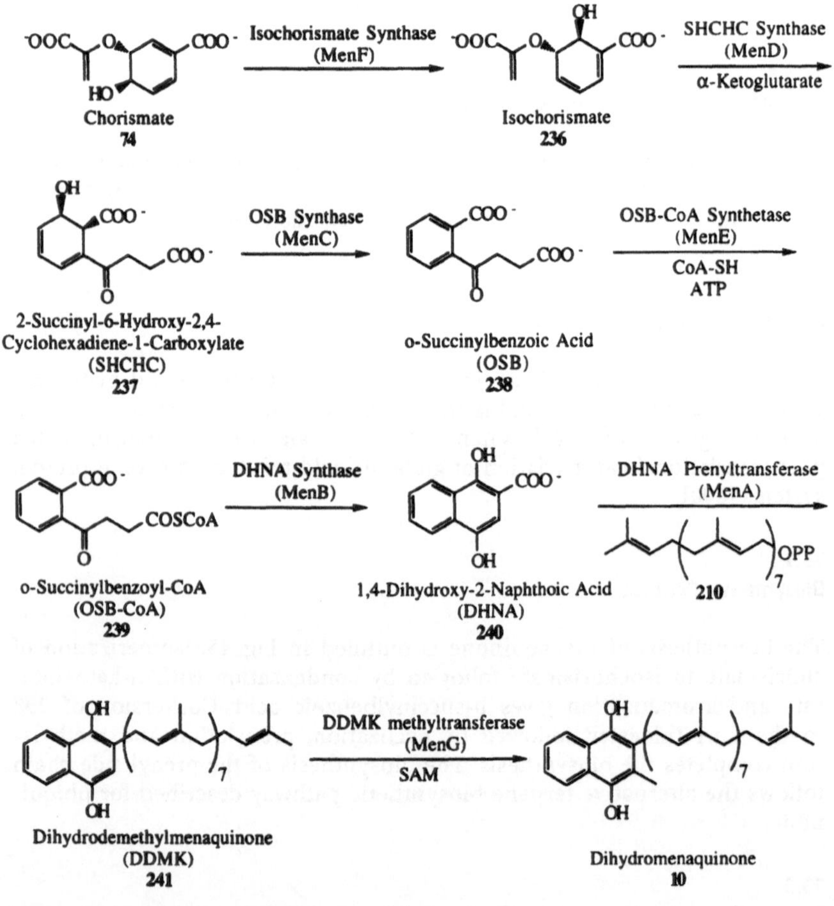

Fig. 44. The menaquinone biosynthetic pathway

13.3
Mechanistic Highlights

13.3.1
Isochorismate Synthase

The mechanism for this enzyme has not yet been determined. Three mechanisms have been proposed and are outlined in Fig. 45 [70].

13.3.2
SHCHC Synthase

The mechanism of this reaction has not been studied. A proposal is outlined in Fig. 46 [204].

13.3.3
DHNA Synthase

The stereochemistry of this reaction has been studied. The pro-R proton is initially removed from C-2 and the pro-S proton is removed from C-3 [217]. A mechanistic proposal is outlined in Fig. 47.

13.3.4
DHNA Prenyl Transferase

The mechanism for this reaction has not been studied. A simple electrophilic substitution as outlined in Fig. 48 is likely. The reaction may involve an S_N2 attack by the phenol on the pyrophosphate or proceed via a carbocation.

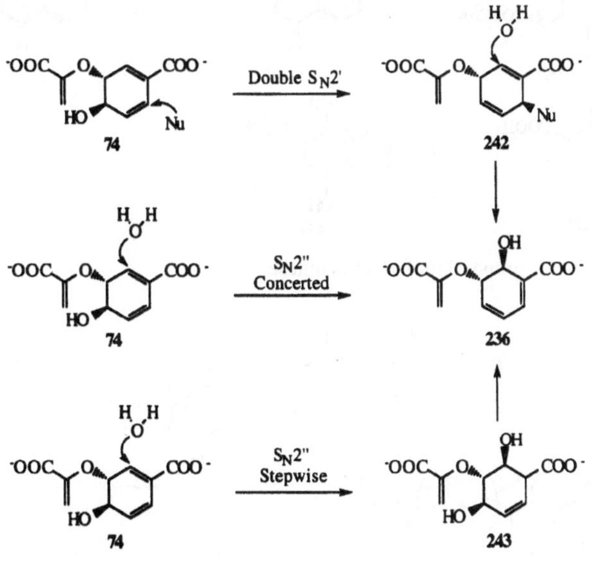

Fig. 45. Mechanistic proposals for isochorismate synthase

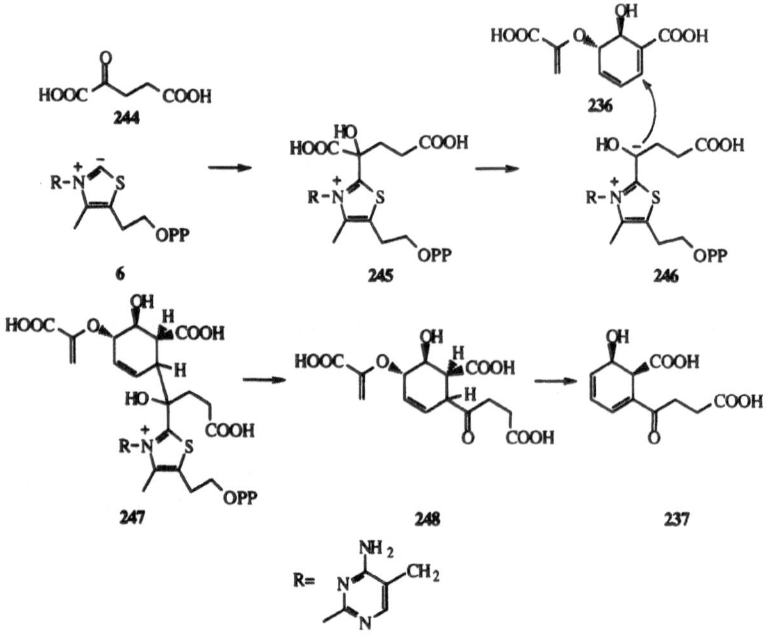

Fig. 46. Mechanistic proposal for SHCHC synthase

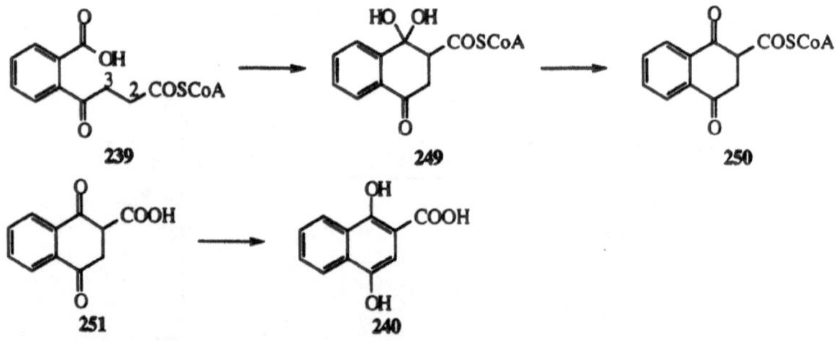

Fig. 47. Mechanistic proposal for DHNA synthase

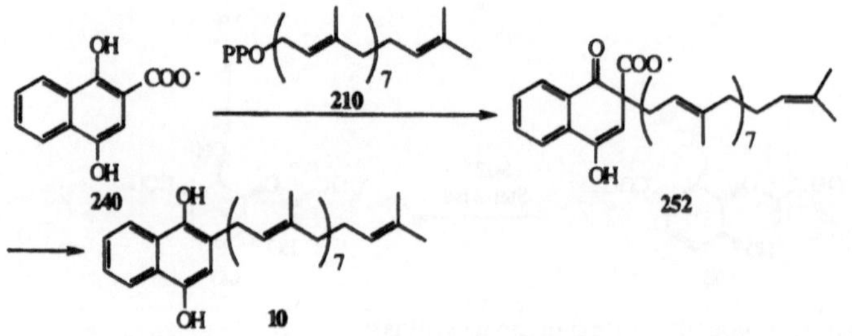

Fig. 48. Mechanistic proposal for DHNA prenyl transferase

14
Summary and Conclusions

After the cofactor biosynthetic pathways had been determined, interest in cofactor biosynthesis declined because the enzymes involved could not be isolated
in quantities suitable for mechanistic studies. This is clearly no longer the case.
Most of the genes involved in the biosynthesis of the *E. coli* cofactors have been
cloned and sequenced but the overexpression and the chemistry lags far behind
the molecular biology. There are now many interesting tractable problems in this
area awaiting solution by chemists interested in solving mechanistic puzzles.

15
References

1. Neidhardt FC, Curtiss III R, Ingraham JL, Lin ECC, Low KB, Magasanik B, Reznikoff WS,
   Riley M, Schaechter M, Umbarger HE (eds) *Escherichica coli* and *Salmonella typhimurium*: Cellular and Molecular Biology. Vol 1. ASM, Washington DC. This is an excellent
   review of cofactor biosynthesis, written from a molecular biology perspective
2. Adams MJ (1987) Oxido-reductases-pyridine nucleotide-dependent enzymes. In: Page
   MI, Williams A (eds) Enzyme Mechanisms. Royal Society of Chemistry, London, p 477
3. Penfound T, Foster JW (1996) Biosynthesis and recycling of NAD. In: Neidhardt FC,
   Curtiss III R, Ingraham JL, Lin ECC, Low KB, Magasanik B, Reznikoff WS, Riley M,
   Schaechter M, Umbarger HE (eds) *Escherichia coli* and *Salmonella typhimurium*: Cellular
   and Molecular Biology. Vol 1. ASM, Washington DC, p 721
4. Enzymes from other sources will be described only if the corresponding enzyme from
   E. coli has not been isolated or studied from a mechanistic perspective.
5. Seifert J, Kunz N, Flachmann R, Läufer A, Jany KD, Gassen HG (1990) Biol Chem Hoppe-
   Seyler 371:239
6. Flachmann R, Kunz N, Seifert J, Gutlich M, Wientjes F-J, Läufer A, Gassen HG (1988) Eur J
   Biochem 175:221
7. Nasu S, Wicks FD, Gholson RK (1982) J Biol Chem 257:626
8. Gardner PR, Fridovich I (1991) Arch Biochem Biophys 284:106
9. Bhatia R, Calvo KC (1996) Arch Biochem Biophys 325:270
10. Willison JC, Tissot G (1994) J Bacteriol 176:3400
11. Zalkin H (1985) Methods Enzymol 113:297
12. Cheng W, Roth JR (1994) J Bacteriol 176:4260
13. Silverman RB (1995) Acc Chem Res 28:335
14. Walpole CSJ, Wrigglesworth R (1987) Oxido-reductases – Flavoenzymes. In: Page MI,
   Williams A (eds) Enzyme Mechanisms. Royal Society of Chemistry, London, p 506
15. Bacher A, Eberhardt S, Richter G (1996) Biosynthesis of riboflavin. In: Neidhardt FC,
   Curtis III R, Ingraham JL, Lin ECC, Low KB, Magasanik B, Reznikoff WS, Riley M,
   Schaechter M, Umbarger HE (eds) *Escherichia coli* and *Salmonella typhimurium*: Cellular
   and Molecular Biology. Vol 1. ASM, Washington DC, p 657
16. Bacher A, Eisenreich W, Kis K, Ladenstein R, Richter G, Scheuring J, Weinkauf S (1993) Biosynthesis of flavins. In: Dugas H, Schmidtchen FP (eds) Bioorganic Chemistry Frontiers.
   Vol 3. Springer, Berlin Heidelberg New York, p 147
17. Perkins JB, Pero J (1993) Biosynthesis of riboflavin, biotin, folic acid and cobalamin. In:
   Sonenshein AL, Hoch JA, Losick R (eds) *Bacillus subtilis* and other gram positive bacteria.
   ASM, Washington DC, p 319
18. Mironov VN, Kraev AS, Chikindas ML, Chernov BK, Stepanov AI, Skryabin KG (1994) Mol
   Gen Genet 242:201
19. Richter G, Ritz H, Katzenmeier G, Volk R, Kohnle A, Lottspeich F, Allendorf D, Bacher A
   (1993) J Bacteriol 175:4045

20. Richter G, Fischer M, Krieger C, Eberhardt S, Luttgen H, Gerstenschlager I, Bacher A (1997) J Bacteriol 179:2022
21. Richter G, Volk R, Krieger C, Lahm H-W, Rothlisberger U, Bacher A (1992) J Bacteriol 174:4050
22. Schott K, Kellermann J, Lottspeich F, Bacher A (1990) J Biol Chem 265:4204
23. Mortel S, Fischer M, Richter G, Tack J, Weinkauf S, Bacher A (1996) J Biol Chem 271:33201
24. Eberhardt S, Richter G, Gimbel W, Werner T, Bacher A (1996) Eur J Biochem 242:712
25. Kearney EB, Goldenberg J, Lipsick J, Perl M (1979) J Biol Chem 254:9551
26. Keller PJ, Van QL, Kim S-U, Blown DH, Chef H-C, Kohnle A, Bacher A, Floss HG (1988) Biochem 27:1117
27. Crout DHG, Hadfield JA (1987) J Chem Soc Chem Commun 742
28. Volk R, Bacher A (1990) J Biol Chem 265:19479
29. Volk R, Bacher A (1991) J Biol Chem 266:20610
30. Volk R, Bacher A (1988) J Biol Chem 110:3651
31. Kis K, Volk R, Bacher A (1995) Biochem 34:2883
32. Ritsert K, Huber R, Turk D, Ladenstein R, Schmidt-Base K, Bacher A (1995) J Mol Biol 253:151
33. Bacher A, Fischer M, Kis K, Kugelbrey K, Mortl S, Scheuring J, Weinkauf S, Eberhardt S, Schmidt-Base K, Hubert R, Ritsert K, Cushman M, Ladenstein R (1996) Biochem Soc Trans 24:89
34. Kis K, Bacher A (1995) J Biol Chem 270:16788
35. Rowan T, Wood HCS (1963) Proc Chem Soc 72:21
36. Rowan T, Wood HCS (1968) J Chem Soc Sect C Org Chem 452
37. Eschenmoser A, Lowenthal E (1992) Chem Soc Rev 21:1
38. Paterson T, Wood HCS (1969) J Chem Soc Chem Commun 290
39. Beach RL, Plaut GWE (1970) J Am Chem Soc 92:2913
40. Paterson T, Wood HCS (1972) J Chem Soc Perkin Trans I 1051
41. Beach RL, Plaut GWE (1969) Tetrahedron Lett 40:3489
42. Plaut GWE, Beach RL, Aogaichi T (1970) Biochem 9:771
43. Brown DH, Keller PJ, Floss HG, Sedlmaier H, Bacher A (1986) J Org Chem 51:2461
44. Scheuring J, Cushman M, Bacher A (1995) J Org Chem 60:243
45. Cushman M, Patel HH, Scheuring J, Bacher A (1993) J Org Chem 58:4033
46. Cushman M, Patrick DA, Bacher A, Scheuring J (1991) J Org Chem 56:4603
47. Oschkinat H, Schott K, Bacher A (1992) J Biomol NMR 2:19
48. Benkovic SJ, Young M (1987) Folate-dependent enzymes. In: Page MI, Williams A (eds) Enzyme Mechanisms. Royal Society of Chemistry, London, p 429
49. Green JM, Nichols BP, Matthews RG (1996) Folate biosynthesis, reduction, and polyglutamylation. In: Neidhardt FC, Curtiss III R, Ingraham JL, Lin ECC, Low KB, Magasanik B, Reznikoff WS, Riley M, Schaechter M, Umbarger HE (eds) *Escherichia coli* and *Salmonella typhimurium*: Cellular and Molecular Biology. Vol 1. ASM, Washington DC, p 665
50. Katzenmeier G, Schmid C, Bacher A (1990) FEMS Microbiol Lett 66:231
51. Katzenmeier G, Schmid C, Kellermann J, Lottspeich F, Bacher A (1991) Biol Chem Hoppe-Seyler 372:991
52. Ritz H, Keller G, Richter G, Katzenmeier G, Bacher A (1993) J Bacteriol 175:1553
53. Suzuki Y, Brown GM (1974) J Biol Chem 249:2405
54. Mathis JB, Brown GM (1970) J Biol Chem 245:3015
55. Lopez P, Lacks SA (1993) J Bacteriol 175:2214
56. Talarico TL, Ray PH, Dev IK, Merrill BM, Dallas WS (1992) J Bacteriol 174:5971
57. Dallas WS, Gowen JE, Ray PH, Cox MI, Dev IK (1992) J Bacteriol 174:5961
58. Bognar AL, Osborne C, Shane B, Singer SC, Ferone R (1985) J Biol Chem 260:5625
59. Bognar AL, Osborne C, Shane B (1987) J Biol Chem 262:12337
60. Smith DR, Calvo JM (1982) Mol Gen Genet 187:72
61. Smith DR, Calvo JM (1980) Nucl Acids Res 8:2255
62. Roux B, Walsh CT (1992) Biochem 31:6904
63. Ye QZ, Liu J, Walsh CT (1990) Proc Natl Acad Sci USA 87:9391

64. Green JM, Merkel WK, Nichols BP (1992) J Bacteriol 174:5317
65. Green JM, Nichols BP (1991) J Biol Chem 266:12971
66. Nar H, Huber R, Meining W, Schmid C, Weinkauf S, Bacher A (1995) Structure 3:459
67. Nar H, Huber R, Auerbach G, Fischer M, Hosl C, Ritz H, Bracher A, Meining W, Eberhardt S, Bacher A (1995) Proc Natl Acad Sci USA 92:12120
68. Silverman RB (1992) The organic chemistry of drug design and drug action. Academic, San Diego, p 156
69. Tesmer J, Klem T, Deras M, Davisson VJ, Smith JL (1996) Nat Str Biol 3:74
70. Walsh CT, Liu J, Rusnak F, Sakaitani M (1990) Chem Rev 90:1105
71. Anderson KS, Kati WM, Ye QZ, Liu J, Walsh CT, Benesi AJ, Johnson KA (1991) J Am Chem Soc 113:3198
72. Teng CY, Ganem B, Doktor SZ, Nichols BP, Bhatnager RK, Vining LC (1985) J Am Chem Soc 107:5008
73. Roux B, Walsh CT (1993) Biochem 32:3763
74. Viswanathan VK, Green JM, Nichols BP (1995) J Bacteriol 177:5918
75. Epstein DM, Benkovic SJ, Wright PE (1995) Biochem 34:11037
76. Lee H, Reyes VM, Kraut J (1996) Biochem 35:7012
77. Benkovic S, Fierke CA, Naylor AM (1988) Science 239:1105
78. Collison D, Garner CD, Joule JA (1996) Chem Soc Rev 26:25
79. George GN, Garrett RM, Prince RC, Rajagopalan KV (1996) J Am Chem Soc 118:8588
80. Rajagopalan KV (1996) Biosynthesis of the molybdenum cofactor. In: Neidhardt FC, Curtiss III R, Ingraham JL, Lin ECC, Low KB, Magasanik B, Reznikoff WS, Riley M, Schaechter M, Umbarger HE (eds) Escherichia coli and Salmonella typhimurium: Cellular and Molecular Biology. Vol 1. ASM, Washington DC, p 674
81. Pitterle DM, Rajagopalan KV (1993) J Biol Chem 268:13499
82. Zurick TR, Pitterle DM, Rajagopalan KV (1991) FASEB J 5:A470
83. Joshi MS, Johnson JL, Rajagopalan KV (1996) J Bacteriol 178:4310
84. Wuebbens MM, Rajagopalan KV (1995) J Biol Chem 270:1082
85. Pitterle DM, Johnson JL, Rajagopalan KV (1993) J Biol Chem 268:13506
86. Vander Horn P, Backstrom A, Stewart V, Begley TP (1993) J Bacteriol 175:982
87. Haake P (1987) Thiamin-dependent enzymes. In: Page MI, Williams A (eds) Enzyme Mechanisms. Royal Society of Chemistry, London, p 390
88. For reviews on thiamin biosynthesis see: a) Begley TP (1996) Nat Prod Rep 177 b) Estramareix B, David S (1996) New J Chem 20:607 c) White RL, Spenser ID, Biosynthesis of thiamin. In: Neidhardt FC, Curtiss III R, Ingraham JL, Lin ECC, Low KB, Magasanik B, Reznikoff WS, Riley M, Schaechter M, Umbarger HE (eds) Escherichia coli and Salmonella typhimurium: Cellular and Molecular Biology. Vol 1. ASM, Washington DC, p 680
89. David S, Estramareix B, Fischer J-C, Therisod M (1982) J Chem Soc Perkin Trans I, 2131
90. Tazuya K, Yamada K, Nakamura K, Kumaoka H (1987) Biochim Biophys Acta 924:210
91. DeMoll E, Shive W (1985) Biochem Biophys Res Comm 132:217
92. Estramareix B, Therisod M (1972) Biochim Biophys Acta 273:275
93. Bellion E, Kirkley D, Faust J (1976) Biochim Biophys Acta 437:229
94. White R, Rudolph F (1978) Biochim Biophys Acta 542:340
95. Estramareix B, Therisod M (1984) J Am Chem Soc 106:3857
96. Estramareix B, David S (1990) Biochim Biophys Acta 1035:154
97. Estramareix B, David S (1986) Biochem Biophys Res Comm 134:1136
98. Tazuya K, Tanaka M, Morisaki M, Yamada K, Kumaoka H (1987) Biochem Int 14:769
99. Costello CA, Begley TP (unpublished results)
100. An alternative pathway to the pyrimidine moiety of thiamin has been discovered in Salmonella typhimurium. a) Downs DM, Roth JR (1991) J Bacteriol 173:6597 b) Downs DM (1992) J Bacteriol 174:1515 c) Petersen L, Downs DM (1996) J Bacteriol 178:5676
101. Backstrom A, McMordie A, Begley TP (1995) J Am Chem Soc 117:2351
102. Backstrom A, Begley TP (unpublished results)
103. Backstrom A, Costello CA, Chiu H-J, Kelleher N, McLafferty F, Taylor S, Begley TP (unpublished results)

104. Webb E, Claas C, Downs D (unpublished results), Kinsland C, Begley TP (unpublished results)
105. Mizote T, Tsuda M, Nakazawa T, Nakayama H (1996) Microbiol 142:2969
106. Nakayama H (1990) Bitamin 64:619
107. Zhang Y, Taylor SV, Chiu H-J, Begley TP (1997) J Bacteriol 179:3030
108. Petersen L, Downs D, (unpublished results)
109. Kawasaki T, Iwashima A, Nose Y (1969) J Biochem 65:407
110. Kawasaki T, Iwashima A, Nose Y (1969) J Biochem 65:417
111. Imamura N, Nakayama H (1982) J Bacteriol 151:708
112. Nakayama H, Hayashi R (1972) J Bacteriol 109:936
113. Webb E, Downs D, (unpublished results)
114. Begley TP (1996) Nat Prod Rep 177
115. DeMoll E (1996) Biosynthesis of biotin and lipoic acid. In: Neidhardt FC, Curtiss III R, Ingraham JL, Lin ECC, Low KB, Magasanik B, Reznikoff WS, Riley M, Schaechter M, Umbarger HE (eds) *Escherichia coli* and *Salmonella typhimurium*: Cellular and Molecular Biology. Vol 1. ASM, Washington DC, p 704
116. Otsuka AJ, Buoncristiani MR, Howard PK, Flamm J, Johnson C, Yamamoto R, Uchida K, Cook C, Ruppert J, Matsuzaki J (1988) J Biol Chem 263:19577
117. Lemoine Y, Wach A, Jeltsch J-M (1996) Mol Microbiol 19:639
118. Ploux O, Marquet A (1996) Eur J Biochem 236:301
119. Ploux O, Marquet A (1992) Biochem J 283:327
120. Stoner GL, Eisenberg MA (1975) J Biol Chem 250:4029
121. Alexeev D, Bury SM, Boys CWG, Turner MA, Sawyer L, Ramsey AJ, Baxter HC, Baxter RL (1994) J Mol Biol 235:774
122. Sanyal I, Cohen G, Flint DH (1994) Biochem 33:3625
123. Ifuku O, Miyaoka H, Koga N, Kishimoto J, Haze S-I, Wachi Y, Kajiwara M (1994) Eur J Biochem 220:585
124. Sanyal I, Lee S-L, Flint DH (1994) J Am Chem Soc 116:2637
125. Ploux O, Soularue P, Marquet A, Gloeckler R, Lemoine Y (1992) Biochem J 287:685
126. Emery VC, Akhtar M (1987) Pyridoxal phosphate dependent enzymes. In: Page MI, Williams A (eds) Enzyme Mechanisms. Royal Society of Chemistry, London, p 343
127. Stoner GL, Eisenberg MA (1975) J Biol Chem 250:4037
128. Gibson KJ, Lorimer GH, Rendina AR, Taylor WS, Cohen G, Gatenby AA, Payne WG, Roe DC, Lockett BA, Nudelman A, Marcovici D, Nachum A, Wexler BA, Marsilii EL, Turner IM, Howe LD, Kalbach CE, Chi H (1995) Biochem 34:10976
129. Huang W, Jia J, Gibson KJ, Taylor WS, Rendina AR, Schneider G, Lindquist Y (1995) Biochem 34:10985
130. Alexeev D, Baxter RL, Smekal O, Sawyer L (1995) Structure 3:1207
131. Sanyal I, Cohen GS, Flint DH (1994) Biochem 33:3625
132. Sanyal I, Gibson KJ, Flint DH (1996) Arch Biochem Biophys 326:48
133. Méjean A, Tse Sum Bui B, Florentin D, Ploux O, Izumi Y, Marquet A (1995) Biochem Biophys Res Commun 217:1231
134. Birch OM, Fuhrmann M, Shaw NM (1995) J Biol Chem 270:19158
135. Parry RJ (1983) Tetrahedron 39:1215
136. Marti FB (1983) Diss ETH (Zurich) Nr 7236
137. Baxter RL, Camp DJ, Coutts A, Shaw N (1992) J Chem Soc Perkin Trans I 255
138. Marquet A, Frappier F, Guillerm G, Azoulay M, Florentin D, Tabet J-C (1993) J Am Chem Soc 115:2139
139. Florentin D, Tse Sum Bui B, Marquet A, Ohshiro T, Izumi Y (1994) CR Acad Sci Paris 317:485
140. Jaun B, Pfaltz A (1988) J Chem Soc Chem Commun 293
141. Saeva FD, Morgan BP (1984) J Am Chem Soc 106:4121
142. Beak P, Sullivan TA (1982) J Am Chem Soc 104:4450
143. Franz JA, Roberts DH, Ferris KF (1987) J Org Chem 52:2256
144. Parry RJ (1977) J Am Chem Soc 99:6464

145. Hayden MA, Huang I, Bussiere DE, Ashley GW (1992) J Biol Chem 267:9512
146. Reed KE, Cronan Jr JE (1993) J Bacteriol 175:1325
147. White RH (1980) J Am Chem Soc 102:6605
148. Hayden MA, Huang IY, Iliopoulos G, Orozco M, Ashley GW (1993) Biochem 32:3778
149. White RH (1980) Biochem 19:15
150. White RH (1986) Anal Chem Symp Ser XI 543
151. Jackowski S (1996) Biosynthesis of pantothenic acid and coenzyme A. In: Neidhardt FC, Curtiss III R, Ingraham JL, Lin ECC, Low KB, Magasanik B, Reznikoff WS, Riley M, Schaechter M, Umbarger HE (eds) Escherichia coli and Salmonella typhimurium: Cellular and Molecular Biology. Vol 1. ASM, Washington DC, p 687
152. Jones CE, Brook JM, Buck D, Abell C, Smith AG (1993) J Bacteriol 175:2125
153. Wilkin DR, King Jr HL, Dyar RE (1975) J Biol Chem 250:2311
154. Shimizu S, Kataoka M, Chung MC-M, Yamada H (1988) J Biol Chem 263:12077
155. Williamson JM (1985) Methods Enzymol 113:589
156. Miyatake KY, Nakano Y, Kitaoka S (1979) Methods Enzymol 62:215
157. Song W-J, Jackowski S (1994) J Biol Chem 269:27051
158. Brown GM (1959) J Biol Chem 234:370
159. Yang H, Abeles RH (1987) Biochem 26:4076
160. Martin DP, Drueckhammer DG (1993) Biochem Biophys Res Commun 192:1155
161. Worrall DM, Tubbs PK (1983) Biochem J 215:153
162. Powers SG, Snell EE (1979) Methods Enzymol 62:204
163. Aberhart DJ (1979) J Am Chem Soc 101:1354
164. Van Poelje PD, Snell EE (1990) Ann Rev Biochem 59:29
165. Scandurra R, Politi L, Santoro L, Consalvi V (1987) FEBS Lett 212:79
166. Aberhart DJ, Ghoshal PK, Cotting J-A, Russell DJ (1985) Biochem 24:7178
167. Markham GD, Hafner EW, Tabor CW, Tabor H (1980) J Biol Chem 255:9082
168. Parry RJ, Minta A (1982) J Am Chem Soc 104:821
169. Markham GD, Parkin DW, Mentsch F, Schramm VL (1987) J Biol Chem 262:5609
170. Takusagawa F, Kamitori S, Markham GD (1996) Biochem 35:2586
171. Zhao G, Winkler ME (1996) FEMS Microbiol Lett 135:275
172. Himmeldirk KK, Kennedy IA, Hill RE, Spenser ID (1996) Chem Commun 1187
173. Hill RE, Himmeldirk K, Spenser ID (1996) J Biol Chem 271:30426
174. Wolf E, Hill RE, Sayer BG, Spenser ID (1995) J Chem Soc Chem Commun 1339
175. Spenser ID, Hill RE (1995) Nat Prod Rep 555
176. Hill RE, Spencer ID (1996) Biosynthesis of vitamin B6. In: Neidhardt FC, Curtiss III R, Ingraham JL, Lin ECC, Low KB, Magasanik B, Reznikoff WS, Riley M, Schaechter M, Umbarger HE (eds) Escherichia coli and Salmonella typhimurium: Cellular and Molecular Biology. Vol 1. ASM, Washington DC, p 695
177. Zhao G, Pease AJ, Bharani N, Winkler ME (1995) J Bacteriol 177:2804
178. Pease AJ, Winkler ME (1993) Abstract K-116. In: Abstracts of the 93rd General Meeting of the American Society for Microbiology. ASM, Washington DC, p 280
179. Man T-K, Zhao G, Winkler ME (1996) J Bacteriol 178:2445
180. a) Zhao G, Winkler ME (1995) J Bacteriol 177:883. b) Notheis C, Drewke C, Leistner E (1995) Biochim Biophys Acta 1247:265
181. a) Yang Y, Zhao G, Winkler ME (1996) FEMS Microbiol Lett 141:89, b) White RS, Dempsey WB (1970) Biochem 9:4057
182. Meganathan R (1996) Biosynthesis of the isoprenoid quinones menaquinone (vitamin K2) and ubiquinone (coenzyme Q). In: Neidhardt FC, Curtiss III R, Ingraham JL, Lin ECC, Low KB, Magasanik B, Reznikoff WS, Riley M, Schaechter M, Umbarger HE (eds) Escherichia coli and Salmonella typhimurium: Cellular and Molecular Biology. Vol 1. ASM, Washington DC, p 642
183. Rohmer M, Seeman M, Horbach S, Bringer-Meyer S, Sahm H (1996) J Am Chem Soc 118:2564
184. White RH (1996) Biosynthesis of isoprenoids in bacteria. In: Neidhardt FC, Curtiss III R, Ingraham JL, Lin ECC, Low KB, Magasanik B, Reznikoff WS, Riley M, Schaechter M,

Umbarger HE (eds) *Escherichia coli* and *Salmonella typhimurium*: Cellular and Molecular Biology. Vol 1. ASM, Washington DC, p 637

185. Schwender J, Seemann M, Lichtenthaler HK, Rohmer M (1996) Biochem J 316:73
186. Nichols BP, Green JM (1992) J Bacteriol 174:5309
187. Siebert M, Bechthold A, Melzer M, May U, Berger U, Schroder G, Schroder J, Severin K, Heide L (1992) FEBS Lett 307:347
188. Siebert M, Severin K, Heide L (1994) Microbiol 140:897
189. Wu G, Williams HD, Gibson F, Poole RK (1993) J Gen Microbiol 139:1795
190. Melzer M, Heide L (1994) Biochim Biophys Acta 1212:93
191. Suzuki K, Ueda M, Yuasa M, Nakagawa T, Kawamukai M, Matsuda H (1994) Biosci Biotech Biochem 58:1814
192. Wessjohann L, Sontag B (1996) Angew Chem Int Ed 35:1697
193. Leppik RA, Young IG, Gibson F (1976) Biochim Biophys Acta 436:800
194. Knoell H-E (1979) Biochem Biophys Res Commun 91:919
195. Leppik RA, Stroobant P, Shineberg B, Young LC, Gibson P (1976) Biochim Biophys Acta 428:146
196. a) Wu G, Williams HD, Zamanian M, Gibson F, Poole RK (1992) J Gen Microbiol 138:2101.
     b) Hsu AY, Poon WW, Shepherd JA, Myles DC, Clarke CF (1996) Biochem 35:9797
197. Lee PT, Hsu AY, Ha HT, Clarke CF (1997) J Bacteriol 179:1748
198. Alexander K, Young IG (1978) Biochem 17:4745
199. Knoell H-E (1981) FEMS Microbiol Lett 10:63
200. Dowd P, Zheng ZB, Hershline R, Ham SW, Naganathan S, Kerns J (1996) Vitamin K mechanism of action and its bearing on vitamin E. In: Wilcox CS, Hamilton AD (eds) Molecular Design and Bioorganic Calalysis. Kluwer, Dordrecht, Netherlands, p 15
201. Daruwala R, Bhattacharyya DK, Kwon O, Meganathan R (1997) J Bacteriol 179:3133
202. Liu J, Quinn N, Berchtold GA, Walsh CT (1990) Biochem 1929:1417
203. Popp JL (1989) J Bacteriol 171:4349
204. Palaniappan C, Sharma V, Hudspeth MES, Meganathan R (1992) J Bacteriol 174:8111
205. Weische A, Garvert W, Leistner E (1987) Arch Biochem Biophys 256:223
206. Popp JL, Berliner C, Bentley R (1989) Anal Biochem 178:306
207. Sharma V, Meganathan R, Hudspeth MES (1993) J Bacteriol 175:4917
208. Kwon O, Bhattacharyya DK, Meganathan R (1996) J Bacteriol 178:6778
209. Sharma V, Hudspeth MES, Meganathan R (1996) Gene 168:43
210. Sieweke H-J, Leistner E (1991) Z Naturforsch 46c:585
211. Kolkmann R, Leistner E (1987) Z Naturforsch 42c:1207
212. Bryant RW, Bentley R (1976) Biochem 15:4792
213. Sharma V, Suvarna K, Meganathan R, Hudspeth MES (1992) J Bacteriol 174:5057
214. Suvarna K, Meganathan R, Hudspeth MES (1994) Abstract K-159. In: Abstracts of the 94th General Meeting of the American Society for Microbiology. ASM, Washington DC, p 303
215. Shineberg B, Young IG (1976) Biochem 15:2754
216. Suvarna R Unpublished results. Cited in: Meganathan R (1996) Biosynthesis of the isoprenoid quinones menaquinone (vitamin K2) and ubiquinone (coenzyme Q). In: Neidhardt FC, Curtiss III R, Ingraham JL, Lin ECC, Low KB, Magasanik B, Reznikoff WS, Riley M, Schaechter M, Umbarger HE (eds) *Escherichia coli* and *Salmonella typhimurium*: Cellular and Molecular Biology. Vol 1. ASM, Washington DC, p 642
217. Igbavboa U, Leistner E (1990) Eur J Biochem 192:441

Biosynthesis of Vitamin B$_{12}$

Alan R. Battersby · Finian J. Leeper*

University Chemical Laboratory, Lensfield Road, Cambridge CB2 1EW, Great Britain.
E-mail: fjl1@cus.cam.ac.uk

Vitamin B$_{12}$ (cyanocobalamin) is the normal isolated form of coenzyme B$_{12}$ (adenosylcobalamin), a structure of marvellous architecture and amazing biological activity. It belongs to the family of tetrapyrroles which includes inter alia the haems and the chlorophylls. This review begins with a brief overview of the biosynthesis of tetrapyrroles in general but then concentrates on recent research on B$_{12}$ biosynthesis. The first main section reviews the biosynthesis of uro'gen III, the last common precursor of all natural tetrapyrroles, concentrating particularly on the three enzymes, porphobilinogen synthase, hydroxymethylbilane synthase and uro'gen III synthase. Crystal structures are available for the second of these enzymes and a new proposal is presented for its detailed mode of action. The second main section reviews the recent discovery of the complete biosynthetic pathway from uro'gen III to hydrogenobyrinic acid in *Pseudomonas denitrificans*, which has revealed some beautiful and totally unexpected chemistry. A short section then describes the many similarities and some differences in the chemistry used by the micro-aerophilic organism *Propionibacterium shermanii* for the synthesis of coenzyme B$_{12}$ compared with that seen in the aerobic *Ps. denitrificans*. Finally an account is given of the remarkable steps needed to complete the synthesis of the coenzyme in both organisms.

Keywords: Vitamin B$_{12}$; Tetrapyrrole; Porphyrin; Coenzyme B$_{12}$; Corrin.

---

* Corresponding author.

Topics in Current Chemistry, Vol. 195
© Springer Verlag Berlin Heidelberg 1998

List of Symbols and Abbreviations

ALA	5-aminolaevulinic acid
copro'gen	coproporphyrinogen
HMBS	hydroxymethylbilane synthase
Kb	kilobase-pairs (of DNA)
NADH	reduced nicotinamide adenine dinucleotide
NADPH	reduced nicotinamide adenine dinucleotide phosphate
NMR	nuclear magnetic resonance
PBG	porphobilinogen
PLP	pyridoxal phosphate
PMP	pyridoxamine phosphate
proto'gen	protoporphyrinogen
SAM	S-adenosyl-L-methionine
uro'gen	uroporphyrinogen
UROS	uroporphyrinogen III synthase

1
Overview of Tetrapyrrole Biosynthesis

The tetrapyrroles are a group of natural products which include the haems (e.g. haem *b* 1), the chlorophylls (e.g. chlorophyll *a* 2) and the corrinoids (e.g. co-enzyme B$_{12}$ 4), see Scheme 1 [1–6]. In addition to these well-known and wide-spread enzymic cofactors, other tetrapyrroles are used in more restricted cases,

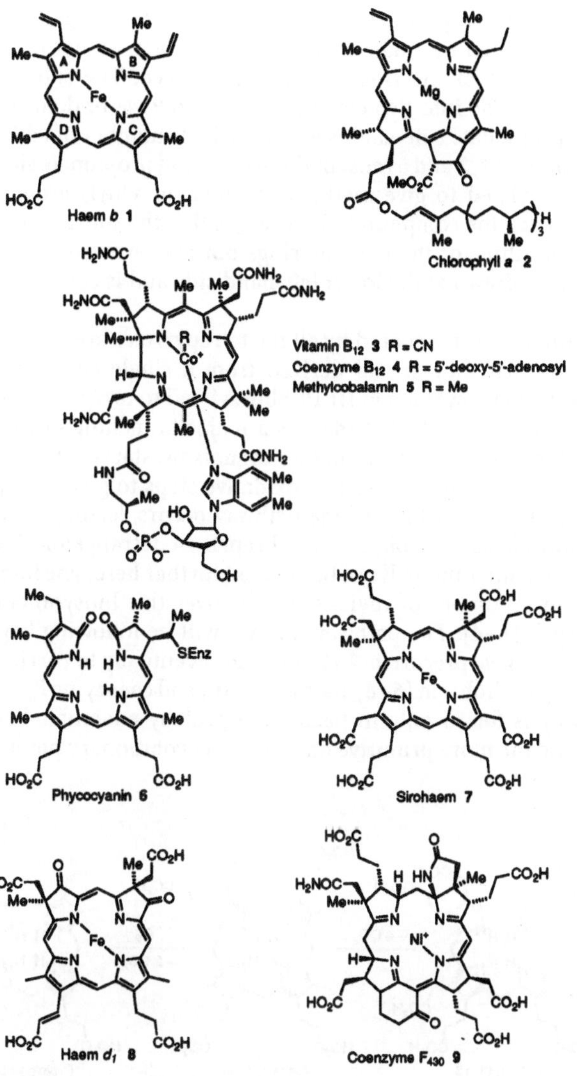

Scheme 1. Some of the more important tetrapyrroles

including bilins (e.g. phycocyanin **6**) used for light-harvesting in algae, siro-haem **7** in sulphite reductase, haem d₁ **8** in the nitrite reductase/cytochrome oxidase of denitrifying bacteria, and coenzyme F₄₃₀ **9** in methyl coenzyme M reductase, the final enzyme in methane production. These compounds are all intensely coloured and every living organism has an absolute requirement for one or more of them. For this reason they have been christened the "pigments of life".

All the natural tetrapyrroles have their four pyrrolic rings, at various oxidation levels, separated by single carbon atom bridges, except in the corrinoids (e.g. 3) where two of the rings are directly linked. The substituents on the pyrrolic rings divide into two classes: in 4, 7 and 9 we find, for the most part, acetate and propionate side-chains with a variable number of additional methyl groups, whereas in 1, 2 and 6 most of the acetate and propionate side-chains have been decarboxylated to give methyl and ethyl (or vinyl) groups. A common feature among all the compounds, however, is that the substituents occur in the same order on three of the pyrrolic rings but are reversed on the fourth one, which is always drawn at the lower left-hand side and is called ring D in all but the bilins.

The common features shared by all the tetrapyrroles are a direct consequence of the fact that they are all derived from a single common tetrapyrrolic macrocycle, uroporphyrinogen III 10, abbreviated to uro'gen III, Scheme 2. It is at the stage of uro'gen III that there is a major branching of the biosynthetic pathways. One branch, found in higher organisms, starts with decarboxylation of uro'gen III and then involves two oxidative steps to give protoporphyrin IX 11. Here a further branching of the pathway occurs, leading eventually to the haems, chlorophylls and bilins. This branch of tetrapyrrole biosynthesis is outlined in Scheme 2 but will not be covered further here. The interested reader is referred to a number of reviews which cover this biosynthetic pathway in some in depth [1–4]. The pathway that we will be following begins with two methylations to give precorrin-2 12 and leads eventually to corrinoids, with off-shoots leading to sirohaem [6–8], haem d_1 [9, 10] and coenzyme F_{430} [11] Scheme 3. This pathway is found only in bacteria, especially anaerobic bacteria, and is believed to be the more primitive one from an evolutionary point of view.

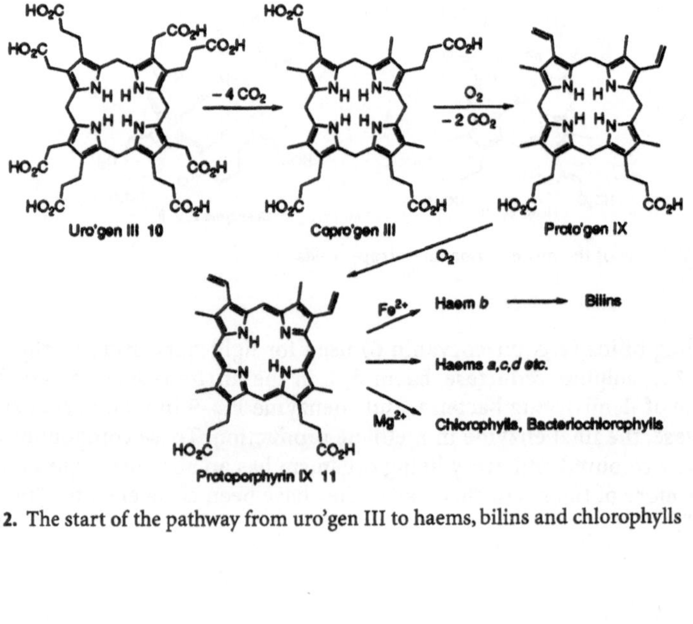

Scheme 2. The start of the pathway from uro'gen III to haems, bilins and chlorophylls

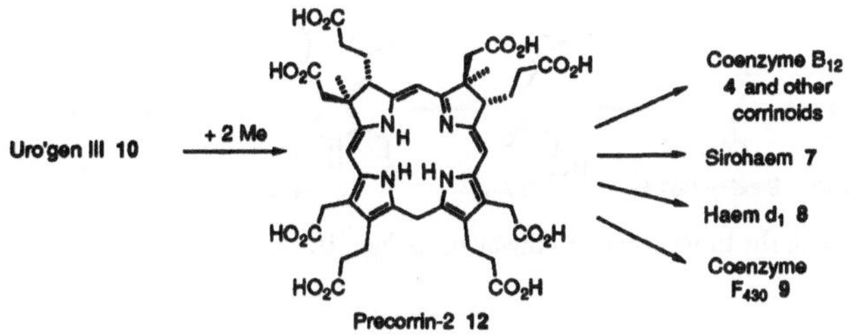

HO₂C

HO₂C

CO₂H

CO₂H

Coenzyme B₁₂ 4 and other corrinoids

Uro'gen III 10 → + 2 Me

Sirohaem 7

Haem d₁ 8

HO₂C

CO₂H

Coenzyme F₄₃₀ 9

HO₂C

CO₂H

Precorrin-2 12

Scheme 3. The start of the B₁₂ pathway from uro'gen III

Vitamin B₁₂ (cyanocobalamin) **3** is, in fact, not a natural product as the cyanide ligand to the cobalt ion is added during the isolation procedure. Coenzyme B₁₂ (adenosylcobalamin) **4** and methylcobalamin **5** are the true final products of the biosynthetic pathway. Coenzyme B₁₂ is the cofactor for a number of enzymic rearrangement reactions, such as that catalysed by methylmalonyl CoA mutase, and methylcobalamin is the cofactor for certain methyl transfer reactions, including the synthesis of methionine. A number of anaerobic bacteria produce related corrinoids in which the dimethylbenzimidazole moiety of the cobalamins (**3–5**) is replaced by other groups which may or may not act as ligands to the cobalt ion, such as adenine or *p*-cresol [12].

2
Biosynthesis of Uroporphyrinogen III

2.1
Introduction

Early experiments on tetrapyrrole biosynthesis in the 1940s and 1950s had established that the pathway shown in Scheme 4 is followed in animals [1–4]. Thus glycine **13** and succinyl CoA **14** are condensed to give 5-aminolaevulinic acid **15** (ALA). Two molecules of ALA are combined to give porphobilinogen **16** (PBG) and then four molecules of PBG are combined to form uro'gen III **10**. Subsequent research has demonstrated that ALA, PBG and uro'gen III are intermediates of tetrapyrrole biosynthesis in all organisms but it has transpired that plants, algae and most bacteria use an alternative route for the production of ALA starting from glutamate.

Once the intermediates had been identified, research concentrated mostly on the individual enzymes of the pathway. The current state of knowledge on each of the enzymes is reviewed in the following sections.

2.2
5-Aminolaevulinic Acid Synthesis

In animals, yeasts and purple photosynthetic bacteria ALA **15** is made from glycine **13** and succinyl CoA **14** by the action of a single enzyme, ALA synthase.

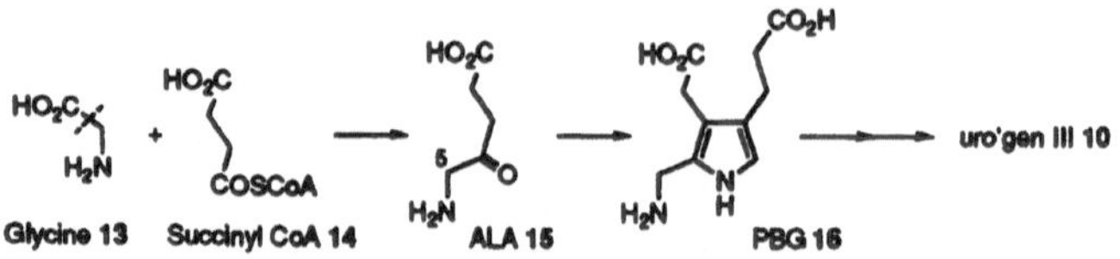

Scheme 4. Outline of the biosynthetic pathway to uro'gen III

This enzyme has been isolated from many organisms and in all cases studied the enzyme is dependent on pyridoxal phosphate (PLP). In isotopic labelling experiments it was found that all four carbon atoms of succinyl CoA **14** are incorporated into ALA along with C-2 and the nitrogen atom of glycine. Furthermore only one of the two hydrogen atoms at C-2 of glycine is incorporated, H_S being lost to the medium while H_S is incorporated into the pro-*S* position at C-5 of ALA [1, 2]. From this information, the mechanism shown in Scheme 5 can be proposed. In order to end up with the observed stereochemistry, either the initial acylation of the glycine occurs with retention of configuration and the subsequent decarboxylation occurs with inversion or vice versa.

From 1974 onwards it became increasingly apparent that in plants a different route to ALA must be involved [1, 2]. It was shown that incorporation of radioactivity from glycine and succinate into chlorophyll *a* **2** was poor compared with the incorporation of five-carbon compounds such as glutamate **17**. ^{14}C labelling of glutamate showed that all five carbon atoms are incorporated and this route to ALA from glutamate is generally known as the C_5 route as opposed to the Shemin route from glycine and succinyl CoA. Subsequently this C_5 route has been shown also to operate in algae and most bacteria, including all the major producers of coenzyme B_{12} and related corrinoids.

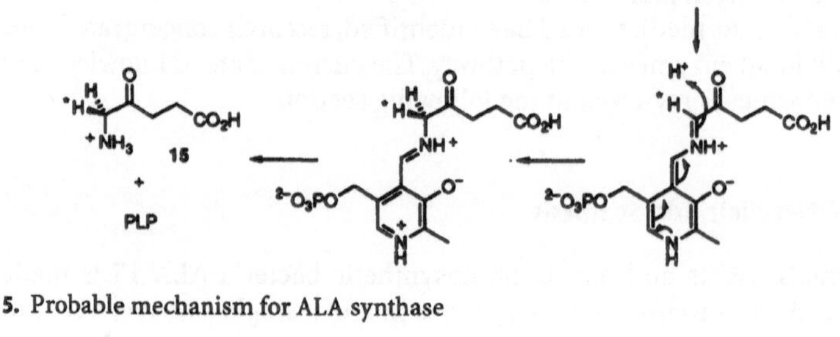

Scheme 5. Probable mechanism for ALA synthase

The individual steps involved in the conversion of glutamate into ALA are shown in Scheme 6 [1, 2, 13]. First the glutamate is converted by an ATP-dependent ligase into a ^{t}RNA ester **18**, which appears to be the same as the glutamyl-^{t}RNA used for protein synthesis in the plant chloroplast. This glutamyl-^{t}RNA is then reduced by an NADPH-dependent reductase to glutamate 1-semialdehyde **19**. As expected of an α-amino aldehyde, **19** is not particularly stable under neutral or basic conditions but can be isolated under acidic conditions, under which it cyclises to the corresponding lactol [2].

The final step is a double transamination reaction which isomerises glutamate 1-semialdehyde **19** to ALA. Logically, this could occur via either PLP-mediated loss of the amino group to give dione **20** followed by regaining of the same amino group at C-5 to give ALA or, alternatively, pyridoxamine phosphate (PMP)-mediated gain of an amino group could give diamine **21** and then loss of the amino group at C-4 would give ALA (Scheme 7). UV spectroscopy indicates

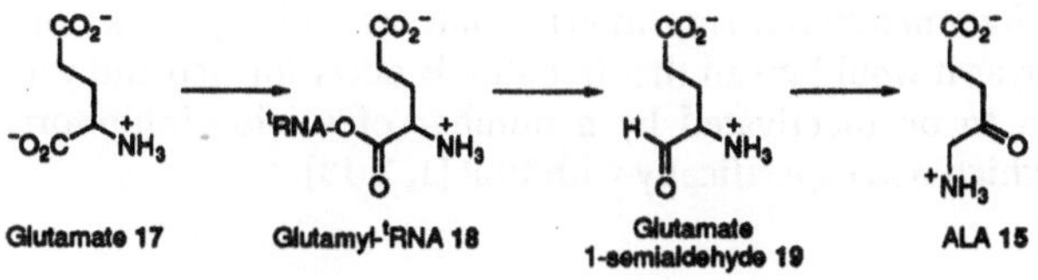

Scheme 6. ALA biosynthesis in plants, algae and most bacteria

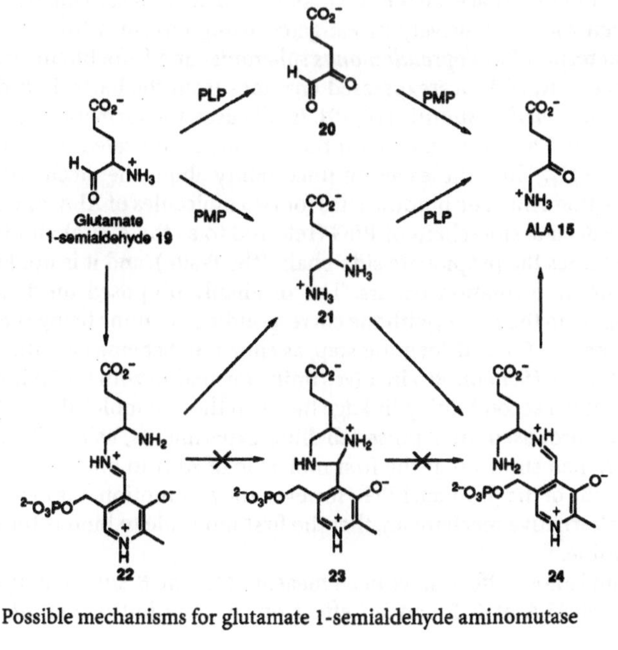

Scheme 7. Possible mechanisms for glutamate 1-semialdehyde aminomutase

that the enzyme isolated from various sources has predominantly PMP bound with a minor amount of PLP [13]. Treatment with the diamine 21 converts the PLP form into the PMP form and stimulates the enzyme, whereas dione 20 increases the concentration of the PLP form. Given these observations, it seems likely that the preferred pathway is via diamine 21. This has been confirmed by a double labelling experiment in which [1-^{13}C]glutamate and [^{15}N]glutamate were mixed and converted into ALA by the enzymes from *Chlamydomonas reinhardtii*. NMR spectroscopy of the resulting ALA revealed that the two labels had come together in some molecules, indicating intermolecular transfer [14]. The reaction sequence proceeding via diamine 21 requires such intermolecular transfer, whereas the alternative sequence via the dione 20 would more likely involve intramolecular transfer of the amino group. Finally, the apoenzyme can only be reactivated by PMP not PLP, again suggesting the pathway via diamine 21 is followed [15].

It is conceivable that instead of hydrolysis of the imine intermediate 22 to give diamine 21 and PLP followed by formation of the alternative imine 24, an intramolecular transfer of the pyridoxyl moiety from one nitrogen to the other occurs via a five-membered ring intermediate 23. This appears not to be the case, however, as it would mean that free PLP is never formed and yet the enzyme is known to be inactivated by a number of suicide inhibitors, such as gabaculine, which react specifically with PLP [1, 2, 13].

2.3
Porphobilinogen Synthase (5-Aminolaevulinic Acid Dehydratase)

The next stage in the biosynthesis, dimerisation of ALA 15 to give PBG 16, is catalysed by PBG synthase, also known as ALA dehydratase. This transformation has been most extensively investigated using enzymes from the photosynthetic bacterium *Rhodopseudomonas spheroides* and from bovine liver [1, 2] and, more recently, with overexpressed enzymes from the bacteria *Escherichia coli* [16, 17] and *Bacillus subtilis* [18, 19]. In all cases, the formation of an imine between ALA and a lysine residue on the enzyme is indicated by inactivation caused by NaBH$_4$. This still leaves an uncertainty about the mechanism, however, because there must be binding sites for two molecules of ALA, one of which provides the acetate side-chain of PBG (referred to as the A-site) and the other of which provides the propionate side-chain (the P-site), and it is not known in which site imine formation occurs. The originally proposed mechanism involved an imine in the A-site, with the corresponding enamine being the nucleophile in the key C–C bond-forming step, as shown in Scheme 8a [20]. An alternative mechanism (Scheme 8b) involves imine formation with the lysine residue in the P-site and a second imine linkage between the two molecules of ALA [21, 22]. This was proposed after pulse-labelling experiments, using ^{13}C- and ^{14}C-labelled ALA, had shown that the first molecule of ALA to bind ends up as the propionate-side of the product, PBG. However, the assumption was made, in proposing the alternative mechanism, that the first molecule to bind is the one that forms the imine.

Deuterium isotope effects have been measured for the *B. subtilis* enzyme [19]: deuteration of ALA at C-5 had no effect on the rate but deuteration at C-3

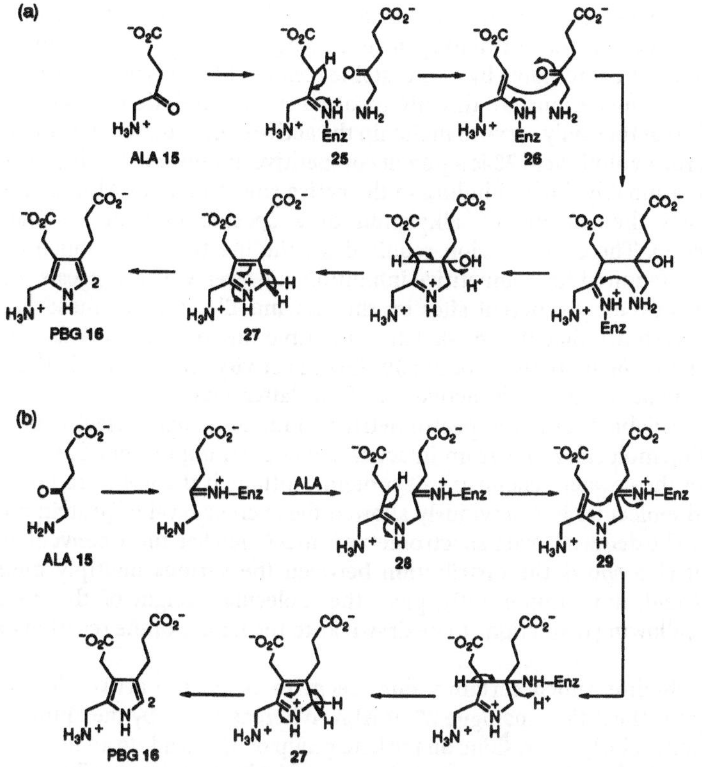

Scheme 8 a, b. Alternative mechanisms for ALA dehydratase

reduced the V_{max} rate by a factor of 3.3 and V_{max}/K_M is 2.3-fold lower. When ALA was reisolated after 50% reaction, no loss of deuterium was detected. These results indicate that the first deprotonation at C-3 of ALA is a rate-determining step and furthermore that the steps leading up to this step must be reversible. The rate-determining deprotonation step would be $25 \rightarrow 26$ in the first mechanism (Scheme 8a) or $28 \rightarrow 29$ in the second mechanism (Scheme 8b).

The stereochemistry of the final deprotonation, at the carbon which becomes C-2 of PBG, has been determined using [5S-³H]ALA (derived from [2-³H]glycine by the ALA synthase reaction, see Sect. 2.2) [1, 2]. The tritium label is entirely retained in the PBG produced, whereas 50% is lost from ALA which is randomly tritiated at C-5. Thus it is the pro-R hydrogen atom that is lost as illustrated in Scheme 8, $27 \rightarrow 16$.

PBG synthase is a metal-requiring enzyme but the metals required vary from one source to another. The bovine enzyme requires zinc, the *E. coli* enzyme binds both zinc and magnesium [16, 17], whereas plant enzymes only require magnesium. In the bovine enzyme two zinc atoms bind per subunit. An EXAFS experiment has shown that the first zinc ion, required for activity, binds in a site where it is five-coordinate, with two or three histidine nitrogen atoms, a sulphur from cysteine and one or two oxygen atom as ligands. The second ion binds in a site where it is coordinated by four cysteine residues [23]. It is not clear, however, whether the metal ions are directly involved in the catalytic mechanism; it is possible that they only serve to maintain the active conformation of the enzyme.

Chlorolaevulinic acid 28 is a potent competitive inhibitor of bovine PBG synthase, presumably due to binding in the active site in place of ALA, and it also inactivates the enzyme by alkylation of a specific cysteine residue [24] (Scheme 9). The concentration required for the inactivation is much greater than that required for competitive inhibition, however, which suggests that the processes occur at different sites on the enzyme. Electrospray mass spectrometry has shown that 28 can alkylate at multiple sites on the B. subtilis enzyme without causing more than about 50% loss of activity [18]. It is likely that there is no cysteine residue in the active site of this latter enzyme.

The availability of mass spectrometric techniques (such as electrospray) for observing molecular ions from intact enzymes is an important advance in the study of the covalent chemistry of proteins. Although the use of radioactively labelled molecules has previously allowed the average level of protein modification to be deduced, mass spectrometry is much quicker, more convenient and safer; it also shows the distribution between the various multiply modified species and, most importantly, gives the molecular weight of the modified species, allowing deductions to be drawn as to the nature of the reaction occurring.

An inhibitor which certainly does seem to act at the active site of PBG synthase is the 3-thia analogue 29 of ALA. Whereas attack by the amino group of the active site lysine residue on the keto group of ALA leads to an imine, attack on the equivalent thioester carbonyl group of 29 leads to irreversible acylation of the amine, as shown in Scheme 9 [18, 25]. ^{13}C NMR was used to demonstrate this formation of an amide linkage and mass spectrometry was again used to demonstrate that a single acylation is sufficient to inactivate the enzyme. Importantly, the rate of inactivation was proportional to the square of the concentration of thioester 29, indicating that two molecules bind before the attack of the amino group occurs. Because of the close similarity of the inactivator and the substrate, it is probable that two molecules of the substrate also bind before imine formation occurs.

It will be obvious from the above that there is still considerable uncertainty about the mechanism of PBG synthase. However there have been reports of crystallisation of the bovine [26] and yeast enzymes [27], and preliminary X-ray

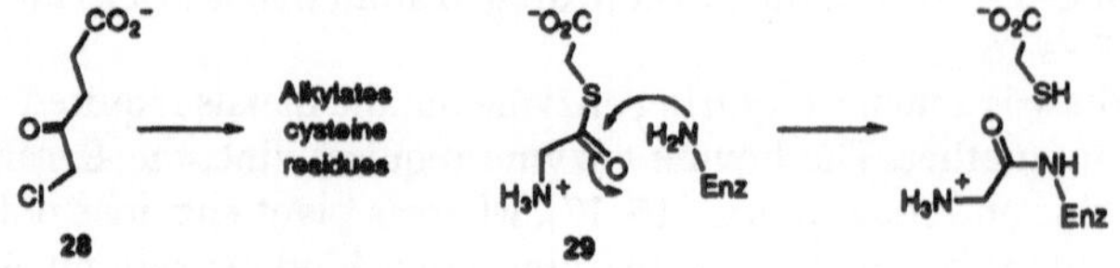

Scheme 9. Inactivation of ALA dehydratase

diffraction data on the former have been obtained, so it is hoped the picture will become much clearer in the near future when the crystal structures have been solved.

2.4
Hydroxymethylbilane Synthase (Porphobilinogen Deaminase)

It was recognised in the 1950s that the next two enzymes in the pathway catalyse the tetramerisation of PBG, with loss of ammonia, to give uro'gen III 10. Heat treatment inactivated one of the two components resulting in the formation of the unrearranged uro'gen I 31 instead of uro'gen III, in which ring D is inverted (Scheme 10). The first enzyme became known as PBG deaminase and the second one was called cosynthetase because it was thought that it somehow interacts with PBG deaminase to modify the product that it forms. It was not until 1978, that it was discovered that in fact PBG deaminase produces hydroxymethyl-bilane 30, which is the true substrate for cosynthetase. In the absence of cosynthetase this intermediate undergoes a rapid non-enzymic transformation into uro'gen I, thus explaining the previous observations [1, 2, 28]. Following this discovery the names hydroxymethylbilane synthase (HMBS) and uro'gen III synthase (UROS) were officially adopted for the two enzymes [29].

It was known from several different experiments that the successive PBG molecules become covalently linked to HMBS during the course of the reaction. However, attempts to identify the enzymic group to which the first PBG molecule became attached were all unsuccessful until two important techniques were introduced: firstly the gene for HMBS in E. coli (hemC) was cloned and over-expressed allowing the production of large quantities of the enzyme and secondly it was found that the enzyme-PBG complexes could be separated on a multimilligram scale by fast protein liquid chromatography (FPLC). These techniques made a range of experiments possible which could not be contemplated before. In the first of these experiments, HMBS was incubated with a slight

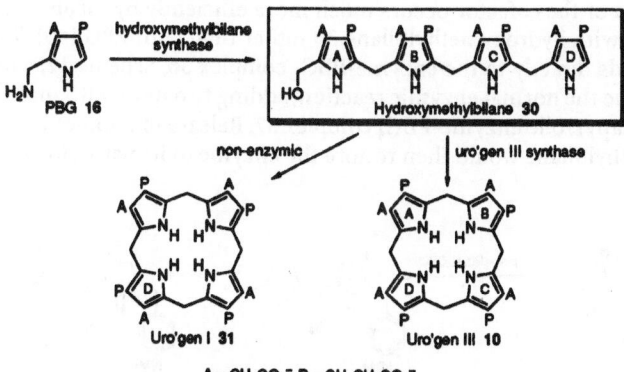

A = CH₂CO₂⁻, P = CH₂CH₂CO₂⁻

Scheme 10. The relationship between hydroxymethylbilane synthase and uro'gen III synthase

excess of [11-^{13}C]PBG and the resulting mixture of complexes was separated to give approximately 10 mg of pure monoPBG complex. The ^{13}C NMR spectrum of this complex, at pH 12 to partially denature the protein, showed a strong new peak at δ 24.6 ppm [30, 31]. From extensive model studies, this chemical shift was known not to correspond to any of the expected types of attachment, i.e. pyrrole-CH$_2$-O, pyrrole-CH$_2$-N or pyrrole-CH$_2$-S [32], but instead it corresponded exactly to pyrrole-CH$_2$-pyrrole. It follows that there must already be an α-free pyrrole on the native enzyme to which the first PBG molecule becomes attached.

The presence of an α-free pyrrole attached to the native HMBS was also demonstrated, both in Cambridge and Southampton, by treatment with Ehrlich's reagent, acidic p-dimethylaminobenzaldehyde [30, 33]. This initially gave the UV/visible absorbance at 564 nm, typical of the Ehrlich product from an α-free pyrrole, but the spectrum then changed to one at 495 nm, typical of a dipyrromethene, indicating that the cofactor is in fact a dipyrromethane (e. g. 32), as shown in Scheme 11 and tautomerisation of the initial product 33 occurs to give 34.

The discovery of the dipyrromethane cofactor was clearly a great step forward in our understanding of the mechanism of HMBS. It did, however, beg the question: to what is the cofactor attached? This question was answered by NMR spectroscopy on ^{13}C-labelled cofactor. Two methods were found to label the cofactor: treatment of the native enzyme with acid cleaved off the cofactor and incubation of the resulting apoenzyme with [11-^{13}C]PBG then regenerated the cofactor in labelled form [34]; alternatively, growing the producing organism in the presence of labelled ALA also produced the cofactor in labelled form[35–37]. These labelling experiments allowed the structure to be unambiguously defined as 32, consisting of two molecules of PBG coupled together and attached to the enzyme via a thioether linkage, Scheme 12. Protein degradation studies in Cambridge [38] and Southampton [36] and site-directed mutagenesis experiments in Texas [39] all identified the site of attachment as cysteine-242 and the way the cofactor is involved in the stepwise building of hydroxymethylbilane 30 is shown in Scheme 12. More recently it has been demonstrated that regeneration of the cofactor occurs much more efficiently by incubation of the apoenzyme with hydroxymethylbilane 30 rather than with PBG [40]. This presumably leads directly to the enzyme-PBG$_2$ complex 36, Scheme 12, which can then continue the normal enzymic reaction, adding two more PBG molecules to give the hexapyrrolic enzyme-PBG$_4$ complex 37. Release of another molecule of hydroxymethylbilane would then restore the enzyme to its native form 32.

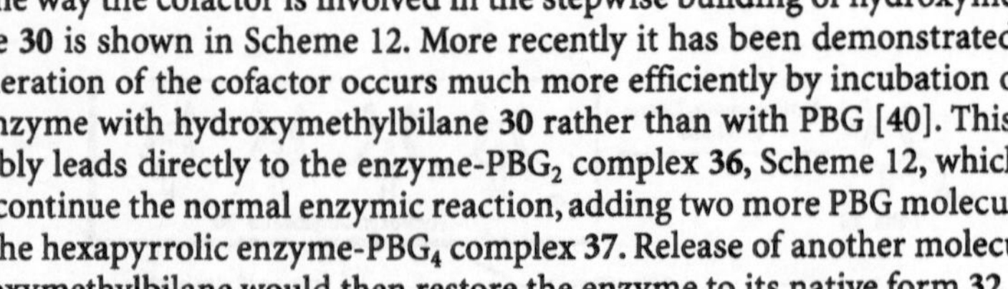

Scheme 11. Reaction of the dipyrromethane cofactor with Ehrlich's reagent

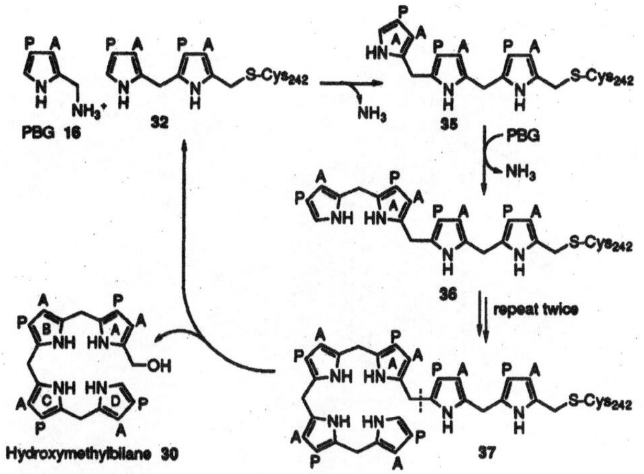

Scheme 12. The mechanism of hydroxymethylbilane synthase

Subsequent to the work just described, all the details were fully confirmed and many more were revealed by X-ray crystal structures of HMBS, Fig. 1 [41 – 44]. The crystal structures show an enzyme consisting of three domains; between domains 1 and 2 is a cavity partly occupied by the dipyrrole cofactor, which is attached to cysteine-242 in the third domain. The nearly planar structure of the dipyrrole in the initially reported crystal structure [41] made it apparent, however, that the dipyrromethane had become oxidised during handling to a species in which the two pyrroles were part of a single conjugated π-system (probably a mixture of dipyrromethene and dipyrromethenone). A second crystal structure was subsequently obtained of the enzyme with the natural reduced cofactor [42, 43]. In this latter structure the first ring of the cofactor is in a very similar position to that of the oxidised cofactor but the second ring is tilted back into the cavity with an angle of 61° between the planes of the two rings. Most recently the structure of the selenomethionine-substituted enzyme with the active reduced cofactor has been solved totally independently using multiwavelength anomalous dispersion (MAD). It is this structure that is shown in Fig. 1.

Significant features of the enzyme revealed by these crystal structures are that the carboxylate group of aspartate-84 is hydrogen bonded to both pyrrolic N-H groups of the cofactor and all but one of the carboxyl groups of the cofactor, in both its reduced and oxidised forms, find good ion pairs with the guanidinium groups of arginine residues. The one exception is the propionate side-chain of ring 2 of the reduced cofactor, which lacks any direct contact with a positively charged counter-ion but does form a water-mediated hydrogen bond with a more distant arginine residue, Arg-176 [45]. This may well be very significant in the mechanism of the reaction, as will be described below.

The importance of all the active-site groups mentioned above has been tested by site-directed mutagenesis. Before the crystal structure was available, a num-

Fig. 1. Ribbon representation of hydroxymethylbilane synthase from *E. coli* [44]. Domains 1, 2 and 3 are in *yellow, blue and magenta* respectively. The cofactor, in *green*, is here in its active reduced form

ber of arginine residues had been changed to either histidine [46] or leucine [47] residues. Mutation of arginine residues 131 and 132, which bind ring 1 of the cofactor, prevents cofactor assembly and so gives inactive enzyme. Mutations of arginines 11, 149, 155 and 176 do not prevent assembly of the cofactor but either prevent or markedly inhibit the later steps of covalent attachment of PBG. Interestingly, the Arg155Leu mutant appears to be unable to release the tetrapyrrole product and so accumulates the enzyme–PBG$_4$ complex, which is not observed with wild-type *E. coli* HMBS. Mutation of other arginine residues (e.g. Arg-101 and 232) have much less effect on the enzymic activity.

Other residues that have been mutated are cysteine-242, aspartate-84, lysine-55 and lysine-59. Changing cysteine-242, the site of attachment of the cofactor, to a serine residue removes essentially all of the catalytic activity by preventing cofactor assembly. When aspartate-84 was changed to alanine or asparagine, all activity was lost, though assembly of the cofactor did still occur [48]. Changing aspartate-84 to a glutamate (i.e. addition of just one CH_2) causes loss of 99% of the enzymic activity. The crystal structure of this mutant shows that the carboxylate group of the glutamate residue does not hydrogen bond to the pyrrole-NH of the second ring of the cofactor, which presumably accounts for its low activity [45]. Lysines 55 and 59 had been shown to be in or near the active site by chemical modification studies using PLP and NaBH$_4$ [49]. Replacement of lysine-55 by glutamine has almost no effect on the enzymic activity, however, whereas replacement of lysine-59 has no effect on k$_{cat}$ but increases the K$_M$ for PBG approximately 30-fold [50], indicating that it plays some part in binding the substrate but not in the catalytic reaction. In the crystal structure, lysine-55 is in a region of the protein which could not be located, presumably because it is mobile or disordered and lysine-59 is just on the edge of this region and its side-chain was not located. It is possible that this region is a loop that closes over the active-site when the substrate is bound.

It seems from the crystal structures that there are three binding sites for pyrrolic rings. The reduced cofactor occupies two of them (sites C1 and C2) and it is likely that the third one, which is occupied by ring 2 of the oxidised cofactor, would normally be the substrate binding site (site S). The substrate must bind the other way round from the orientation observed for ring 2 of the oxidised cofactor, however, because the aminomethyl carbon, C-11 of PBG, must become bonded to C-9 at the end of the dipyrromethane cofactor. With this restriction in mind, we have modelled the binding of the substrate in this binding site using molecular mechanics to predict the energies of various different orientations [51]. The lowest energy orientation, shown in Fig. 2, has a number of interesting

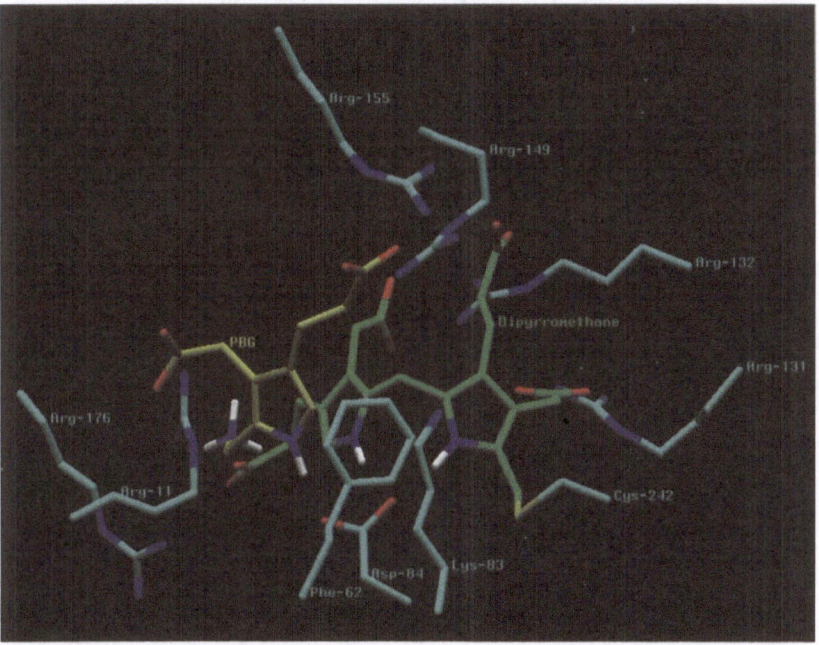

Fig. 2. Close up of the active site of HMBS with the first substrate molecule (PBG) in its proposed non-covalently bound orientation

features. The pyrrole ring of the substrate is significantly further forward than ring 2 of the oxidised cofactor and is also turned sideways so that, instead of having a face-to-face interaction with the phenyl ring of phenylalanine-62, it now has an edge-to-face interaction, with 2-H of PBG pointing towards the centre of the phenyl ring. This edge-to-face interaction probably both favours formation of the intermediate azafulvenium ion **38**, Scheme 13, and protects it from attack by nucleophiles such as water. Despite this change of orientation compared to ring 2 of the oxidised cofactor, the side-chains of the bound PBG still make good ion pairs with the same arginine residues (Arg-11 and Arg-149). A most significant feature is that, whereas in solution PBG forms an intramolecular ion pair between its protonated amino group and the carboxylate of the

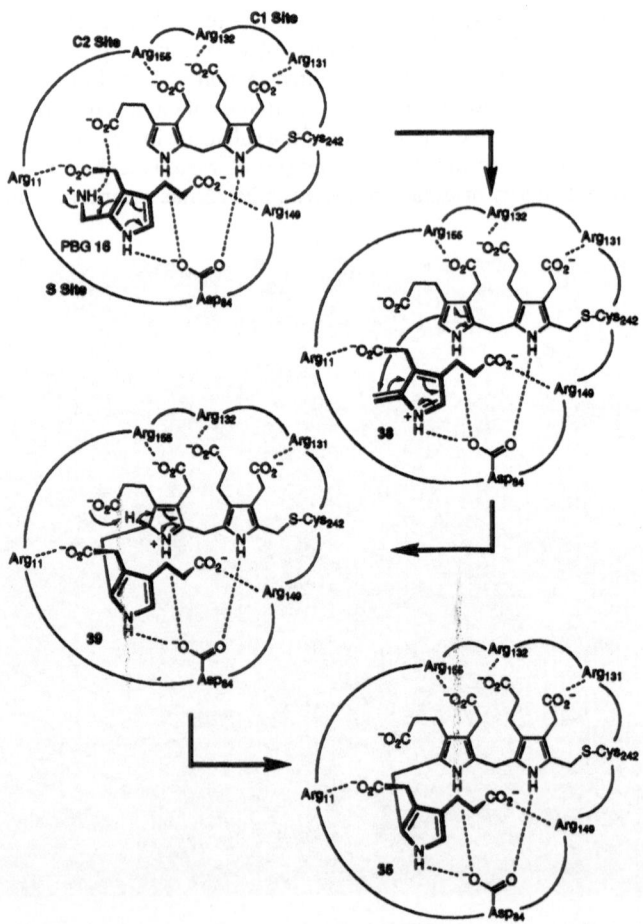

Scheme 13. Proposed details of the mechanism for the first cycle of HMBS

neighbouring acetate side-chain [52], in its predicted enzyme-bound orientation the acetate carboxylate swings forwards to ion-pair with arginine-11 and the protonated amine swings backwards to ion-pair with the carboxylate of the propionate side-chain on ring 2 of the reduced cofactor. This immediately explains why this carboxylate lacks a direct ion-pair in the native enzyme and why the protein forces this side-chain to adopt a gauche conformation rather than the preferred anti conformation (see Fig. 2). The protonated amino group is now nearly perpendicular to the plane of the pyrrole ring, in the correct position to allow expulsion of ammonia to form the azafulvenium ion 38, Scheme 13. Carbon-carbon bond formation can then take place to give intermediate 39, in which ring 2 of the cofactor is a protonated pyrrolenine. Deprotonation at C-9 of the cofactor must then take place to restore this ring to being a pyrrole, as in 35. However, inspection of the crystal structure reveals no enzymic residue capable of effecting this deprotonation. Instead the carboxylate group of the propionate side-chain is perfectly situated to remove the proton from C-9 and having lost its counterion when ammonia was released, it would be a suitably basic group.

The steps shown in Scheme 13 complete the first cycle of covalent attachment of PBG to the dipyrromethane cofactor, giving the tripyrrole 35. This cycle now has to be repeated three more times. Given the excellent orientation of the binding and catalytic groups in the three pyrrolic binding sites identified above and the lack of any further such groups nearby, it seems likely that the whole pyrrolic chain moves along so that the second ring of the cofactor occupies site C1, previously occupied by the first ring of the cofactor, and the now covalently attached substrate ring occupies site C2, previously occupied by the second ring of the cofactor. This would then vacate the substrate binding site ready for the next substrate to enter. This movement is probably made favourable by the strain generated as the central ring becomes a planar pyrrole in the deprotonation step 39 → 35. It is conceivable that the movement of the growing pyrrolic chain is made possible by a conformational change of the protein in which domain 3 swings away from domains 1 and 2 (see also below). There is evidence that such a conformational change does take place because the native enzyme is relatively unreactive towards thiol-modifying reagents such as methyl methanethiosulphonate [53] and N-ethylmaleimide [54] but it becomes increasingly more susceptible to inactivation as increasing numbers of substrate molecules become bound. The residue which is modified when this inactivation occurs has been identified as cysteine-134, which is located on domain 2 at the interface with domain 3 [54].

After four PBG molecules have been attached, giving hexapyrrole 37, Scheme 12, they must be detached to produce the hydroxymethylbilane 30 and regenerate the dipyrrolic cofactor 32. Mechanistically, this is the reverse of the very first attachment step and it presumably occurs in the same site. In other words the two cofactor rings return to their original binding sites, C1 and C2, and ring A occupies the substrate binding site, S. Protonation at C-9 on cofactor ring 2 then occurs (equivalent to 35 → 39) followed by C–C bond cleavage (equivalent to 39 → 38) and capture of the resulting azafulvene by water. If we assume that this mechanism is the exact reverse of the first attachment step then the stereochemistry at the methylene group involved will be the same, i.e. either both occur with retention of configuration (as predicted in Scheme 13) or both occur with inversion. Either way, the resulting overall stereochemistry at the hydroxymethyl carbon of hydroxymethylbilane 30 would be retention of configuration. This has been proved experimentally by incubation of [11R-^{3}H] and [11S-^{3}H]PBG with HMBS followed by chemical degradation of the hydroxymethylbilane produced to glycolic acid and enzymic analysis [55]. It is known, from enzymic conversion of the hydroxymethylbilane through to protoporphyrin IX 11, that the reactions forming each of the three interpyrrolic methylene groups are also stereospecific [1, 2, 28]. However the stereochemistries of these steps have not yet been discovered.

It would be of great interest to obtain crystal structures of the HMBS-substrate complexes so that the nature of the conformational change that occurs as the polypyrrole chain grows can be observed directly. Unfortunately these complexes, while stable enough to be separated by chromatography, are not indefinitely stable and break down by release of monopyrrole units during the crystallisation process. A promising alternative approach, however, is to diffuse PBG into the crystals and then obtain crystal structures at various time intervals using the Laue method, which involves a very brief irradiation from an intense synchrotron X-ray source. This technique is currently being applied to HMBS to study the build-up of the enzyme-substrate complexes and the preliminary results are extremely encouraging [56]. For example it is clear that if domain 3 does move, the shift is not sufficient in the early stages to disrupt the crystal lattice. Also the indications are that aspartate-84 does not significantly change position as the PBG is bound, which fits in with the foregoing mechanistic scheme.

Because it has not yet been possible to crystallise the complexes of HMBS with the natural substrate, there has been considerable interest in analogues of PBG, which might form more stable complexes. A number of analogues have been tested in which PBG has altered side-chains [57]. In quite a few cases the substrate analogues can bind covalently to the cofactor but cannot complete the reaction. An example of this is 2-bromoPBG 40, Scheme 14, which becomes covalently attached to the native enzyme, enzyme-PBG and enzyme-PBG$_2$ complexes and blocks any further reaction, resulting in effectively irreversible inhibition [39, 58]. The 2-fluoro-11-hydroxy analogue 41 of PBG behaves similarly [59] – a hydroxyl group at C-11 is known to be displaced almost as easily as the normal amino group, making 42 a good substrate for HMBS [60].

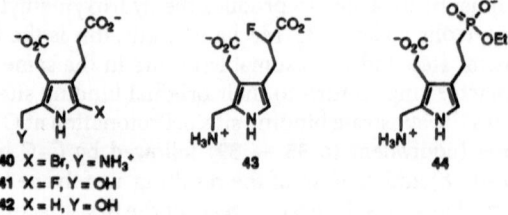

40 X = Br, Y = NH$_3$$^+$
41 X = F, Y = OH
42 X = H, Y = OH

Scheme 14. Some inhibitors and unnatural substrates of HMBS

Of all the analogues of PBG that have been tested, only two, 9-fluoroPBG **43** and the phosphonate analogue **44**, have been shown to be able to complete the whole reaction, producing a modified hydroxymethylbilane [61]. The turnover rates for these two compounds as sole substrates were approximately 100 and 500 times slower respectively than the rate for PBG (6 min⁻¹). For the fluoroPBG a K_M value of 108 μM was measured. When 9-fluoroPBG was tested as an inhibitor of the normal turnover of PBG, an apparent K_I value of 6 μM was measured. This is surprisingly low because normally the K_M value for a slow substrate is the same as its K_I value as a competitive inhibitor. Closer inspection of the kinetic data revealed that the graph of rate vs [PBG] in the presence of the inhibitor is not the usual hyperbolic curve but is sigmoid at low [PBG]. The reason for this and for the low apparent K_I seems to be that after one molecule of fluoroPBG has covalently bound to the enzyme, the binding of the next molecule of PBG (or of fluoroPBG) is affected and a higher concentration is required to achieve half-maximal rate.

The fact that the maximum turnover rate for 9-fluoroPBG is so much lower than for PBG is also something of a surprise because at first sight it would seem that the modification is a minor one and is well removed from the part of the molecule where the reaction occurs. However, if the mechanistic hypothesis presented in Scheme 13 is correct, then this propionate side-chain will play a vital role in the deprotonation step, analogous to **39** → **35**, in the subsequent round of addition of a substrate molecule. The introduction of a fluorine atom α to a carboxyl group lowers its pK_a substantially (e.g. pK_a for CH_3CO_2H is 4.76 and for FCH_2CO_2H is 2.59) and so the carboxylate anion is a weaker base and the deprotonation step would be slower. The involvement of this propionate side-chain both in binding the next substrate molecule and as the base for a key step in the mechanism would also explain why modified PBG molecules in which the length of this side-chain is altered or the carboxyl group is missing are generally able to bind covalently to the enzyme but the complexes so formed can only proceed with attachment of a further pyrrole very slowly if at all [62, 63].

2.5
Uroporphyrinogen III Synthase (Cosynthetase)

The earliest knowledge of the nature of the rearrangement catalysed by uro'gen III synthase (UROS) came with the incorporation of [2,11-¹³C₂]PBG **16a** (derived from [5-¹³C]ALA) diluted with unlabelled PBG into uro'gen III **10a** and thence into protoporphyrin **11a** [1, 2, 28]. The pattern of labelling observed by ¹³C NMR analysis is shown in Scheme 15a. The key point to note is that two ¹³C labels from the same PBG molecule become adjacent at C-15 and C-16, with no mixing with the unlabelled PBG. This shows that at some point during the process the pyrrole ring that provides ring D has been detached from its methylene group, C-11, turned round, and reattached by an intramolecular process. If uro'gen III is derived from [5-¹³C]ALA without dilution with unlabelled ALA, then all the sites marked in **10a** will be ¹³C-labelled. Thus in the ¹³C NMR spectrum all the enriched peaks appear as doublets except for C-15 which is a triplet and C-20 which is a singlet. This signature has been used many times to

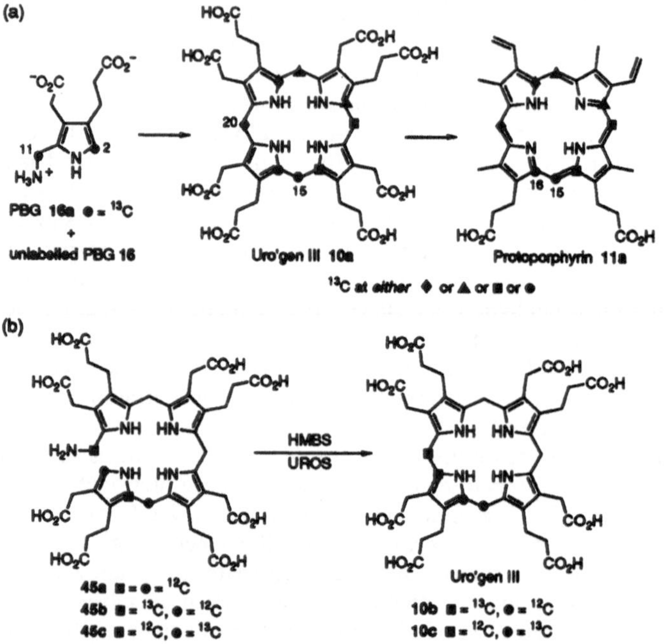

Scheme 15 a, b. Labelling experiments to define the nature of the rearrangement catalysed by UROS

identify tetrapyrrolic compounds derived from uro'gen III and to assist in the assignment of the ^{13}C NMR signals.

The same conclusion with regard to the rearrangement can be drawn from two experiments using labelled aminomethylbilanes **45b** and **45c**, Scheme 15b, but in addition these experiments showed that (i) the rearrangement occurs at the bilane stage and (ii) it occurs intramolecularly within the bilane. The aminomethylbilane **45a** is a product formed by HMBS when high concentrations of ammonia are present. It is not a substrate for UROS but is converted by HMBS into hydroxymethylbilane **30**, the normal substrate for UROS.

A number of modified hydroxymethylbilanes have been synthesised and tested as substrates for UROS [64–66]. These experiments showed that if the bilane is accepted as a substrate, the resulting porphyrinogen usually has ring D predominantly rearranged but a lesser amount is cyclised enzymically without rearrangement. Even in the case of the hydroxymethylbilane having ring D already rearranged this ring is rearranged back again by the enzyme 45 % of the time, giving uro'gen I, whereas 55 % of the time the enzymic cyclisation occurs without rearrangement to give uro'gen III [64].

A likely mechanism for turning round ring D involves reaction of some electrophile at C-16 of the hydroxymethylbilane. This would then permit a fragmentation of the C-15/C-16 bond, allowing ring D to be turned over and reattached via its alternative α-position. In principle the electrophile could be a proton or an iminium ion. However, the hydroxymethyl carbon of the substrate can generate a good electrophile and a bond needs to be made between this carbon and C-16 at some point during the mechanism. Therefore the most economical mechanism is that shown in Scheme 16. This has been termed the "spiro mechanism" because its key intermediate is the spiropyrrolenine **46**.

When the spiro mechanism was first proposed in 1961 [67], it was thought that the macrocyclic portion of the spiro intermediate **46** would be too strained and it was suggested that each pyrrole ring would have to be protonated on one of its α-carbons to engender greater flexibility in the macrocycle. In order to test whether this proposal is necessary, it was decided to aim to synthesise compounds having the same tripyrrolic macrocycle. Several examples of this type of macrocycle were made in Cambridge, all of which proved to be stable compounds [68]. Of the compounds synthesised, the one closest in structure to the spiropyrrolenine **46** was spirolactam **48** [69], which differs only in the replacement of the imine by an amide in the five-membered ring (Scheme 17). This spirolactam proved to be a potent inhibitor of UROS, binding about ten times more strongly than the substrate. The inhibition by spirolactam **48** strongly suggests that a closely related structure (i.e. spiropyrrolenine **46**) is involved in the mechanism.

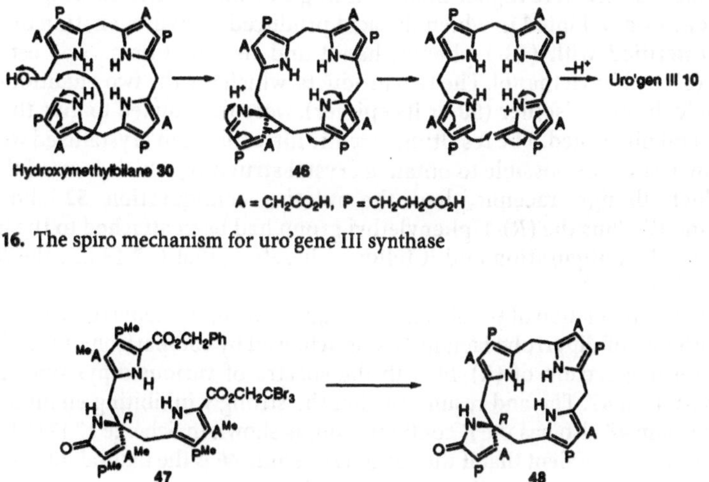

A = CH_2CO_2H, P = $CH_2CH_2CO_2H$

Scheme 16. The spiro mechanism for uro'gene III synthase

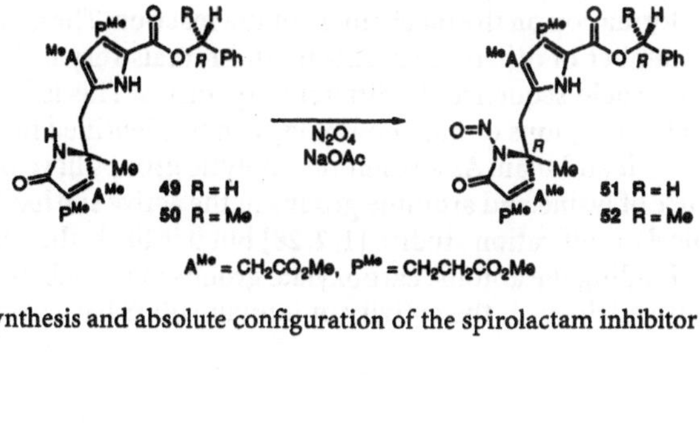

A^{Me} = CH_2CO_2Me, P^{Me} = $CH_2CH_2CO_2Me$

Scheme 17. Synthesis and absolute configuration of the spirolactam inhibitor of UROS

Further evidence for the specificity of the interaction of UROS with the spirolactam **48** came when its two enantiomers were separately prepared by including a resolution step halfway through the synthesis [70]. One of the two enantiomers inhibited UROS at least 20 times more strongly than the other. The unsubstituted parent spiro ring system of **48** does in fact have a plane of symmetry and the difference between the enantiomers lies only in the arrangement of the acetate and propionate side-chains on the three pyrrolic rings of the macrocycle. This makes the discrimination shown by the enzyme between the two enantiomers all the more impressive.

It was not initially possible to assign an absolute configuration to the strongly inhibiting enantiomer of spirolactam **48** and it took a good deal of work over a number of years before this problem was finally solved [71]. A large number of chiral derivatives were prepared related to the optically active dipyrrolyl lactam **47** from which spirolactam **48** had been synthesised. Unfortunately none of these compounds proved to be crystalline and eventually attention turned to monopyrrolyl lactams such as **49**. A derivative that was found to be highly crystalline and gave good X-ray data was the racemic *N*-nitrosolactam **51**. However, the individual enantiomers of this same compound failed to crystallise. It seemed that in this series the chances of crystallising a racemic mixture were higher than for a single enantiomer. Accordingly a novel approach was adopted in which the acid produced by hydrogenation of (–)-**49** was esterified with (*R*)-1-phenylethanol and its enantiomer was esterified with (*S*)-1-phenylethanol. The two products, which are the two enantiomers of a single diastereoisomer (**50** or its epimer), were then mixed to give the racemate and nitrosated. The resulting racemic nitrosolactam crystallised well and at long last it was possible to obtain a crystal structure, which showed that the product, though racemic, had the *relative* configuration **52** shown in Scheme 17. Thus the (*R*)-1-phenylethyl group had been attached to the lactam with the *R* configuration and it follows, therefore, that (–)-**49** has the *R* configuration.

Finally, correlation of the absolute configuration of monopyrrolyl lactam (*R*)-**49** with that of dipyrrolyl lactam **47** was achieved by comparison of the circular dichroism spectrum of (*R*)-**49** with the spectra of various dipyrrolic lactams derived from **47**. The end result was that the strongly inhibiting enantiomer of spirolactam **48** also has the *R* configuration, as shown in Scheme 17 [71]. We can, therefore, be confident that if the spiropyrrolenine **46** is the natural intermediate, it also has the *R* configuration, shown in Scheme 16.

Apart from the inhibition by spirolactam **48**, enzymic studies on UROS have revealed little information on the mechanism of the reaction. The gene sequence is known for a number of different organisms and reveals very little homology between the amino acid sequences in different organisms. This is in contrast to HMBS in which large regions of high homology can be identified for organisms as different as *E. coli* and man. As a result no catalytic groups have been identified. The presence of lysine and arginine groups in the active site has been inferred from chemical modification studies [1, 2, 28] but it is likely that these are the counterions for binding the anionic carboxylate groups of the substrate. A number of mutations which affect the activity of human UROS have been discovered by studying patients with congenital erythropoietic porphyria in whom the enzyme is defective [72].

Proposals for the detailed three-dimensional mechanism of UROS have been put forward [73, 74]. The mechanism really only requires one acidic group to protonate the HO group in the initial step and one basic group to perform the final deprotonation. It is possible that the same enzymic residue could perform both functions. Apart from this, the architecture of the active site must be such that, for the natural substrate **30**, it favours the turning round of ring D and prevents the direct cyclisation to give uro'gen I.

Uro'gen III synthases from most sources are particularly unstable and can be totally inactivated by a brief heat-treatment which leaves HMBS fully active. This instability would frustrate any attempts to crystallise the enzyme. However, it has recently been reported that the UROS from *B. subtilis* is much more stable than from other sources [75] and this opens up the possibility of crystallisation and an X-ray crystal structure determination.

3
Biosynthesis of Hydrogenobyrinic Acid in *Pseudomonas denitrificans*

3.1
Introduction

Progress on the biosynthetic pathway from uro'gen III **10** to vitamin B$_{12}$ **3** and coenzyme B$_{12}$ **4** during the last 5 – 10 years has come almost entirely from research involving the aerobic B$_{12}$-producer *Pseudomonas denitrificans*. There have been dramatic and surprising developments from the work on this organism that have led to the elucidation of the complete biosynthetic pathway to vitamin B$_{12}$. This rapid surge came largely from collaborative research carried out in Paris and in Cambridge. A detailed review of this vast effort has been published [76] and so in this review we will only highlight several key elements that pick out this work as a new departure in the way studies can be made of the biosynthesis of a complex molecule. Also, we will describe in some detail how a selection of the key biosynthetic intermediates were discovered and how their structures were determined on a micro-scale. The research on the other intermediates, not selected for full description, followed very similar lines and these structures are, of course, included in the presentation of the entire pathway to vitamin B$_{12}$ **3** that forms the climax to this chapter, Schemes 29, 30 and 33.

3.2
Earlier Researches Leading to the Recent Advances

Over the past 50 years or so, there has been a transformation in our knowledge of how natural substances are biosynthesised. This was achieved largely by methods based on isotopic labelling and generally involved the following steps: (i) discover what simple building blocks are used and where they fit into the final structure, (ii) from the pointers so gained, postulate likely advanced intermediates on the pathway, (iii) synthesise these putative intermediates in labelled form to test whether they are transformed into the final product by the living system (or an extract from it), (iv) use the clues from (ii) also to search directly for new intermediates by analytical screening of the contents of the organism of interest. It was a very powerful approach that was also used in research on vitamin B_{12} up to around the mid-1980s. The organism selected for that work was a micro-aerophilic one, *Propionibacterium shermanii* and without giving any details of that major effort, it showed that uro'gen III is the parent macrocycle used to construct vitamin B_{12} in addition to the chlorophylls and haems. The step that triggers a branch in the biosynthetic pathway away from chlorophyll and haem towards vitamin B_{12} is *C*-methylation with *S*-adenosyl-L-methionine (SAM) being the methylating agent.

More *C*-methylations follow this first one and the research during this period also set in place the next three intermediates beyond uro'gen III 10 en route to vitamin B_{12} 4, Scheme 18 [1, 28, 77, 78]. They are precorrin-1 53 (or a tautomer of this structure), precorrin-2 12 and precorrin-3A 55; they can be seen as the mono-, di- and tri-methylation products of uro'gen III 10, Scheme 18. The figures after the name "precorrin" correspond to the number of *C*-methyl groups introduced into uro'gen III from SAM to generate that intermediate; the reason for using 3A rather than just 3 will become clear later. When labelled forms of precorrin-2 12 and precorrin-3A 55 were incubated with a crude cell-free enzyme preparation from *Pr. shermanii*, they were converted into cobyrinic acid 58, a late precursor of vitamin B_{12} [1, 28, 77, 78].

The three foregoing intermediates 53, 12 and 55 are all readily oxidised to the corresponding aromatic chlorins or isobacteriochlorins when handled in air. Thus, 53 affords Factor I 54, 12 gives sirohydrochlorin 56 and 55 is converted into Factor III 57. In fact, most of the research leading to the structures of these three precorrins was carried out on the stable dehydrogenated products [1, 28, 77, 78]. Fortunately, the aromatic systems 56 and 57 could be reduced either enzymically or by catalytic hydrogenation to return to precorrin-2 12 and precorrin-3A 55, respectively [1, 28, 77, 78]. It should also be noted that sirohydrochlorin 56 is the metal-free macrocycle corresponding to sirohaem 7 [6–8], Scheme 1.

Despite large efforts to isolate new intermediates from *Pr. shermanii* lying beyond precorrin-3A 55 on the pathway to B_{12}, none were found. Also, following the approaches (i)–(iv) above, some putative later intermediates were prepared in labelled form but were found not to act as precursors of vitamin B_{12}. It was clear that there were simply too many possible ways for the true pathway to go forward from precorrin-3A 55 to make this a realistic approach. A fresh approach was needed and the way the rapid advances over the past decade or so were achieved will be described in the remainder of this chapter.

3.3
Synergistic Combination of Biology, Chemistry and Spectroscopy

Three of the most important changes that were made in the approach to the problem were: (a) to add the power of genetics and molecular biology to the armory of other methods that had previously been used, (b) to use a different organism, the aerobic B_{12}-producer *Pseudomonas denitrificans*, (c) to apply extensive

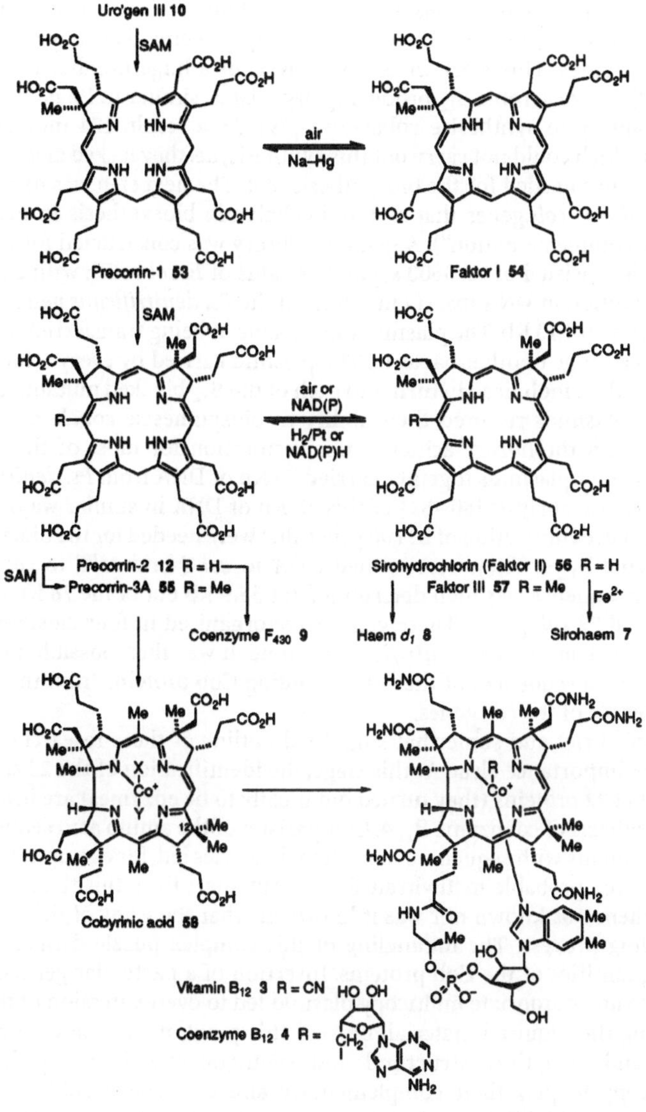

Scheme 18. Earlier knowledge of the pathway to vitamin B$_{12}$ in *Pr. shermanii*

multiple labelling with carbon-13 and then to harness the latest advances in pulsed NMR for determinations of structure.

Step (a) was crucially important in that *now the primary focus was on the enzymes catalysing the various conversions of one intermediate into the next; this then led forward logically to detection and isolation of the biosynthetic intermediates.* The earlier approach had generally been the other way round.

The genes of *Ps. denitrificans* involved in the biosynthesis of coenzyme B_{12} **4** are known as *cob* genes and the corresponding proteins that they encode are the Cob enzymes. This work involved analysis of a huge number of mutants (ca. 22000) of two aerobic B_{12}-producing organisms, similar to *Ps. denitrificans*, for their ability to synthesise cobalamin [79]. As a result, 174 mutants were identified which could not carry out this synthesis, i.e. they lacked at least one of the set of genes needed for the biosynthesis of **4**. The next step was to clone the *Ps. denitrificans cob* genes that restored cobalamin biosynthesis in these Cob mutants ("complementation"). A genomic library was constructed for this purpose which consisted of ca. 3600 separate strains of *E. coli*, each with a plasmid carrying a different DNA insert cut out from the *Ps. denitrificans* genome, with an average size of 13 Kb. The plasmid was capable of being transferred (or mobilised) into a range of other bacteria. The plasmid carried by every strain in the library was then mobilised in turn into each of the B_{12}-blocked mutants above to see which plasmids restored their ability to biosynthesise cobalamin. Eleven plasmids from the library achieved this restoration for most of the blocked mutants. The 11 plasmids together carried 78 Kb of DNA from *Ps. denitrificans*. Further genetic analysis [80–84] of this 78 Kb of DNA in similar ways eventually led to the identification of 22 *cob* genes that were needed for the biosynthesis of coenzyme B_{12} **4**. They were named *cobA* to *cobQ* and *cobS* to *cobW*. The nucleotide sequence was then determined for 35.9 Kb out of the 78 Kb of DNA, including all 22 *cob* genes. These genes were organised in four clusters, somewhat scattered, on the *Ps. denitrificans* genome. It was then possible to predict the amino acid sequences of the corresponding Cob proteins from the nucleotide sequences of the *cob* genes.

Even this brief and somewhat simplified outline of these researches makes their great importance clear. At this stage, the identification of the 22 *cob* genes showed that 22 proteins (they turned out mostly to be enzymes) are involved in the biosynthesis of coenzyme B_{12} **4**. Comparison of the amino acid sequences of the Cob proteins with sequences in protein databases led, for example, to identification of the probable methyltransferases but apart from this the function of none of them was known nor was it known at what stage any of them acted in the building process. The untangling of this complex puzzle demanded substantial quantities of the Cob proteins. Insertion of a particular gene, or set of genes, into an appropriate multicopy plasmid led to overexpression of the gene, so yielding the required material. Beyond this point, it was the turn of enzymology and analytical, structural and synthetic chemistry together with spectroscopy to play their complementary and essential part. The message deserves special highlighting that it was *the combination of genetics, biology and chemistry that led to research on B_{12}-biosynthesis moving forward successfully with a speed and completeness beyond anyone's expectations.*

3.4
First Steps of the Pathway in *Ps. denitrificans*

It was soon found that the early stages of B_{12}-biosynthesis in *Ps. denitrificans* follow those already established in *Pr. shermanii*. Here again, uro'gen III **10** is methylated at C-2 and C-7 to form precorrin-2 **12**, Scheme 18; the double methylation is catalysed by a single enzyme, S-adenosyl-L-methionine:uroporphyrinogen III methyltransferase (SUMT), encoded by the *cobA* gene. This enzyme was purified and characterised [85], as was also that responsible for methylation of **12** at C-20 to form precorrin-3A **55**, named S-adenosyl-L-methionine:precorrin-2 methyltransferase (SP$_2$MT) and encoded by *cobI* [86]. The stereochemistry of the deprotonation steps at C-5, -10 and -20 during the conversion of **10** into **12** has been probed by incorporation of PBG chirally deuterated at C-11 [87]. The result was the same whether (11R) or (11S)-[11-^{2}H]PBG was used: all the deuterium is washed out at both C-5 and C-20, whereas most of the deuterium is retained at C-10. It is likely that the washout occurs by imine–enamine tautomerisation at intermediate stages of the process and that the final deprotonation at C-10 to form precorrin-2 **12** is a non-stereospecific, non-enzymic process which largely retains the deuterium atom due to a normal isotope effect [87].

Over the years there had been no progress in identifying intermediates in *Pr. shermanii* between precorrin-3A **55** and cobyrinic acid **58** and at the outset for *Ps. denitrificans* the long pathway stretching forward from **55** to coenzyme B$_{12}$ **4** was also unmapped. Detection of all the missing intermediates in *Ps. denitrificans* was made in a number of different ways; these will be illustrated by the examples selected for review and in each case the key element or elements of the successful approach will be italicised. Also, the central strategies for the structural and mechanistic research will be picked out in the same way.

3.5
Detection of Precorrin-6A and Elucidation of its Structure

Many recombinant strains of *Ps. denitrificans* were generated in which various *cob* genes had been amplified. One of these overexpressed a complete cluster of the eight *cob* genes F, G, H, I, J, K, L and M and, intriguingly, it produced a substance that showed a strong yellow fluorescence under ultraviolet light. This material was isolated and shown to be hydrogenobyrinic acid **60** [88], Scheme 19, the cobalt-free analogue of cobyrinic acid **58**, Scheme 18. This pro-

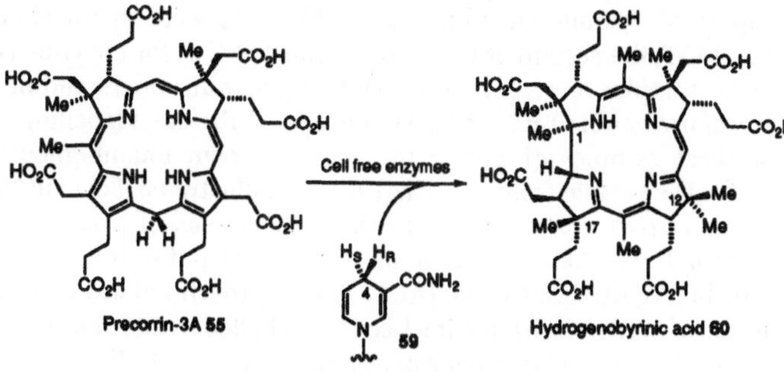

Scheme 19. Biosynthesis of hydrogenobyrinic acid from precorrin-3A

duct was ideal for biosynthetic studies of the B_{12}-pathway in that it represented the complete macrocycle of the vitamin 3 and it had the advantage for handling of being metal-free. Importantly, a cell-free enzyme preparation from the above strain, rich in the eight enzymes encoded by the eight *cob* genes, efficiently transformed precorrin-3A 55 through all the many steps into hydrogenobyrinic acid 60 [89].

Two cofactors were found to be essential for the production of hydrogen-obyrinic acid 60 from precorrin-3A 55, namely SAM, as would be expected, but also reduced nicotinamide adenine dinucleotide phosphate (NADPH, partial structure 59) which was surprising, Scheme 19. *Omission of NADPH from the incubation gave a critically important result; no 60 was formed but a new pale-yellow product appeared in its place.* When a labelled form of this new pigment was incubated with the enzyme system, now with NADPH included, it was specifically converted into hydrogenobyrinic acid 60 in high yield. Clearly, a new intermediate for B_{12}-biosynthesis had been found which opened the door to dramatic progress [89].

Appropriate labelling experiments by the French group [89] showed that production of the new intermediate involved three more methyl groups being added from SAM to precorrin-3A 55. The new substance was thus precorrin-6A and there were strong indications that precorrin-6A had a contracted macrocycle. Further, by labelling the three new methyl groups of precorrin-6A with ^{13}C, it was shown that these appeared at C-17, C-12α and C-1 of the hydrogenobyrinic acid 60 biosynthesised from it, Scheme 19. Mass spectrometry measurements on the methyl ester of precorrin-6A gave two unexpected results: (a) precorrin-6A was found to be an octacarboxylic acid, showing that the C-12 acetate group was still intact, yet at some stage in the biosynthesis of hydrogenobyrinic acid it must undergo decarboxylation to form the 12β-methyl group; (b) precorrin-6A corresponded in oxidation level to a dehydrocorrin (seven double bonds) not to a corrin (six double bonds) and therefore oxidation had occurred before precorrin-6A and there had to be a subsequent reduction to reach the final metal-free corrin 60.

A major difficulty in elucidating the structure of precorrin-6A was that only small amounts (300–500 µg) of non-crystalline material was available from each enzymic run. This was overcome by using ^{13}C-labelled precursors at greater than 90 atom% enrichment to provide samples for analysis by pulsed NMR. Three samples of 5-aminolaevulinic acid, ALA 15, were synthesised, one ^{13}C-labelled at C-5, the second at C-4 and the third at C-3, for enzymic conversion into three samples of precorrin-3A. Drawing on our understanding of the biosynthesis of uro'gen III, reviewed earlier in Sect. 2, the exact labelling pattern of each of these samples of precorrin-3A was known unambiguously; the pattern 55a for the sample from [4-^{13}C]ALA 15a is illustrated, Scheme 20. Each sample was converted enzymically into three specimens of precorrin-6A; in addition, [*methyl*-^{13}C]SAM was used in this last step for the experiments starting with [4-^{13}C]ALA and [3-^{13}C]ALA. This approach led to every atom of the macrocycle of precorrin-6A having been ^{13}C-labelled in one or other of these three final samples. The octamethyl ester of each one was studied by ^{13}C-NMR and by ^{1}H-^{13}C correlation spectroscopy, the latter technique picking out ^{1}H-^{13}C

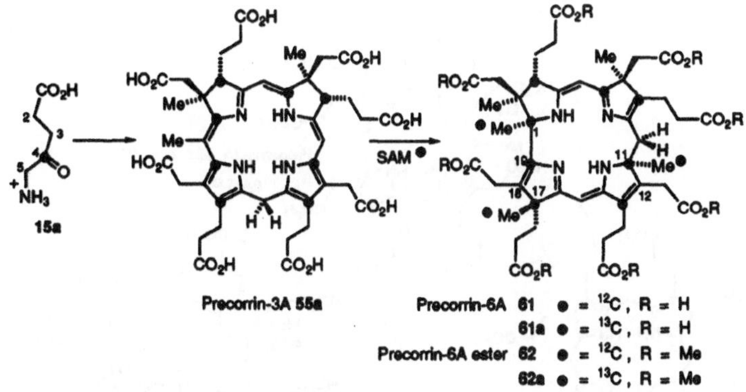

Scheme 20. Example of ¹³C-labelling used for determination of the structure of precorrin-6A

couplings through up to three bonds. *This approach gave information not only about the nature and carbon connectivity of each labelled carbon but also allowed a connectivity pattern of ¹H and ¹³C around the whole macrocycle to be established* [90]. An independent set of connections from measurements of ¹H nuclear Overhauser effects interlocked with that from the ¹H-¹³C couplings.

A full account of these researches cannot be given here but the approach can be briefly illustrated by the experiments based on precorrin-3A **55a** from [4-¹³C]ALA **15a** and using [*methyl*-¹³C]SAM to prepare precorrin-6A **61a**, Scheme 20. The ¹³C-NMR spectrum of the corresponding ester **62a** showed direct connection of C-1 and C-19, thus confirming that ring-contraction had occurred. Also, C-19 was an sp² centre whereas in hydrogenobyrinic acid **60** it is sp³, so indicating that the extra double bond in precorrin-6A **61** that is subsequently reduced is the one at C-18/C-19. Very surprisingly, the NMR signals from the three ¹³C-labelled methyl groups were all doublets showing that all were directly bonded to ¹³C-atoms. The methyl groups at C-1 and C-17 were expected to be so attached, see **60**, but not the third one. This was expected, a priori, to be sited on the unlabelled C-12, where it appears in hydrogenobyrinic acid **60** biosynthesised from precorrin-6A. It was clear that this methyl group must undergo rearrangement to C-12 in **60** from an adjacent site in precorrin-6A and the NMR evidence pointed to methylation at C-11 in the ester **62** [90].

That methylation had occurred at C-11 of precorrin-6A was rigorously proved to be true by non-enzymic synthesis of [11-¹³C]uro'gen III **10d** followed by enzymic conversion into [11-¹³C]precorrin-3A **55d**. Finally, this product was transformed enzymically into precorrin-6A **61d** using [*methyl*-¹³C]SAM, Scheme 21. The ¹³C NMR signal from the C-11 methyl group of the ester **62d** was a doublet as was the signal from C-11 itself, the only ¹³C in the macrocycle; proof of the C-11 methylation was complete [91].

By combining the information gained from all the foregoing experiments with similar data from the studies based on [5-¹³C]ALA and [3-¹³C]ALA, the surprising structure **62**, Scheme 20, was established for precorrin-6A ester [90]; precorrin-6A itself is thus **61**. The configurations at C-2, C-3 and C-7 are based

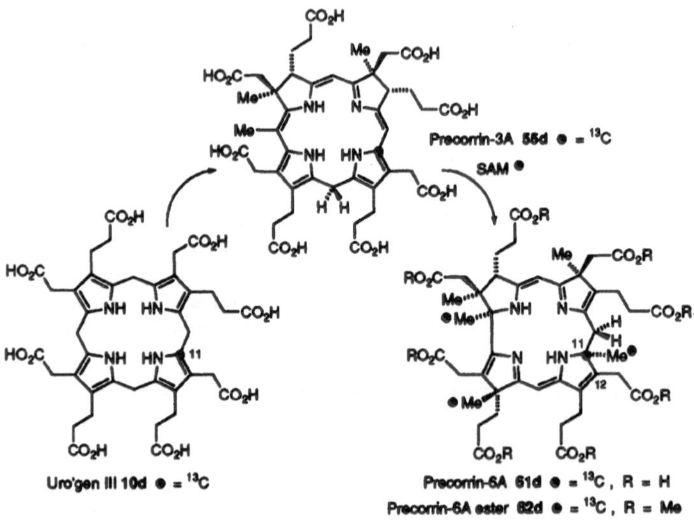

Scheme 21. Proof that precorrin-6A carries a methyl group on C-11

on the efficient enzymic conversion of precorrin-6A **61** into hydrogenobyrinic acid **60**, where the configurations at these centres are firmly established. Also, the *assumption* was made that rearrangement of the methyl group from C-11 of **61**, at some stage, to C-12 of **60** is intramolecular and therefore suprafacial. On this basis, the C-11 methyl group of precorrin-6A **61** was assigned the α-configuration since it is the C-12 α methyl group of the corrin system **58** that is derived from SAM [1, 28, 77, 78]. Experiments will be outlined in Sect. 3.8 that confirm the intramolecular nature of the rearrangement. Note that the C-11 methyl group blocks formation of a fully conjugated system resulting in the pale yellow colour of precorrin-6A.

The structure **61** for precorrin-6A forced a complete change in the way the biosynthesis of vitamin B_{12} was viewed. This structure was full of surprises: (a) the ring contraction step does not occur near the end of the pathway but has been completed at precorrin-6A **61** and is possibly carried out even earlier; (b) oxidation has occurred prior to the formation of precorrin-6A and reduction is needed after it; so the oxidation level along the pathway is not constant; (c) decarboxylation of the C-12 acetate group is not an early step; it occurs on the pathway beyond precorrin-6A; (d) the 12α-methyl group of hydrogenobyrinic acid **60** is introduced initially at C-11 with subsequent migration to C-12.

3.6
Precorrin-6B and the Regio- and Stereospecificity of Precorrin-6A Reductase

Since precorrin-6A **61** had been accumulated by excluding NADPH from the incubation, it was highly probable that precorrin-6A was the substrate for a reductase enzyme which had been rendered inactive by lack of its NADPH co-factor. *Now having all the required enzymes in hand, they could be screened for their ability to transform precorrin-6A.* It turned out that precorrin-6A was indeed enzymically reduced with the addition of two hydrogens to form pre-corrin-6B; the letters A and B are used to distinguish two intermediates having the same level of methylation. Initially, a mixture of enzymes was used for this work in the absence of SAM to avoid further transformation of precorrin-6B [92] but later the pure reductase was isolated by the above screening approach and characterised [92]. Analysis of its amino acid sequence showed that this enzyme is encoded by the *cobK* gene.

The structure determination for precorrin-6B followed the same approach outlined above for precorrin-6A 61 starting from [5-¹³C]ALA and, separately, from [4-¹³C]ALA; the latter experiment is illustrated in Scheme 22. The inter-mediate precorrin-3A 55a yielded precorrin-6B 63a and its octamethyl ester 64a was studied by NMR using the full plethora of techniques already described. As a result, the structure 64 was established for the ester and hence 63 for pre-corrin-6B itself [93]. To expand a little on that statement, C-19 of 63, which is directly bonded to C-1, was clearly an sp³ carbon whereas in precorrin-6A 61 it was sp² hybridised. Thus, the C-18/C-19 double bond of 61 is the one that is re-duced in forming precorrin-6B, which then stands at the same oxidation level as hydrogenobyrinic acid 60.

Now it was possible to answer two further questions: (a) which of C-18 or C-19 receives the hydride equivalent from NADPH and (b) from which face of NADPH 59 is the hydride delivered. *The answer to (a) came from the synthesis of [4-²H₂]NADPH with 95 atom% ²H at C-4 59a,* which was used for the enzymic reduction of precorrin-6A to yield ²H-labelled precorrin-6B. Analysis of this product by ¹H-NMR and mass spectrometry then showed that the ²H had been transferred solely to C-19 [94], Scheme 23. On mechanistic grounds, it seems certain that the transfer of the hydride ion occurs to the C-18 protonated form 65 of precorrin-6A.

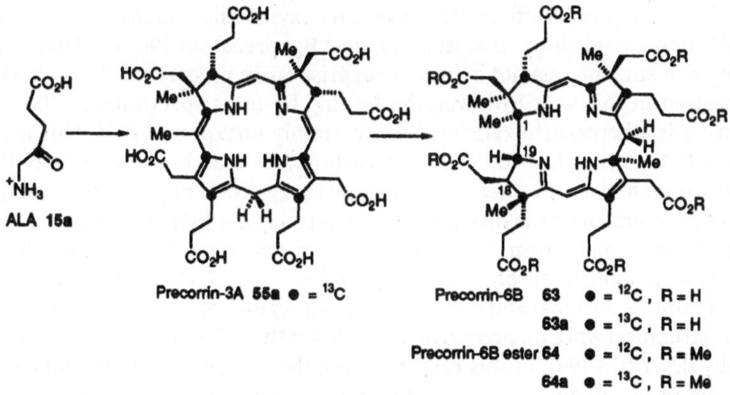

Scheme 22. Structural study of precorrin-6B by ¹³C-labelling

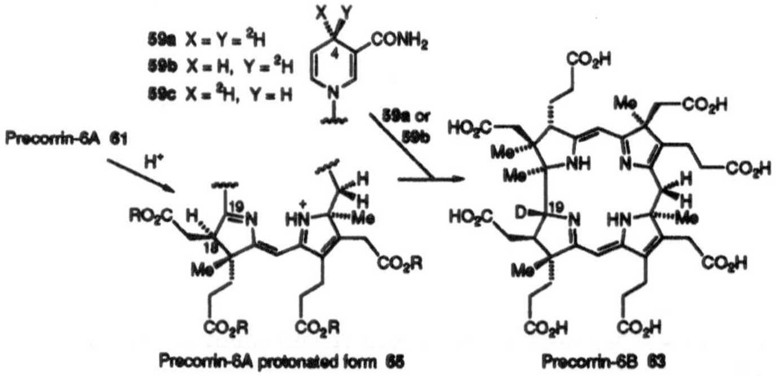

Scheme 23. Stereo- and regio-specificity of reduction of precorrin-6A to precorrin-6B

Question (b) was answered by synthesising [4R-²H]NADPH **59b** *and [4S-²H]NADPH* **59 c**, Scheme 23, which were used in separate experiments to produce two samples of precorrin-6B. NMR analysis as before proved that ²H was only transferred to precorrin-6B **63** from **59b** so the reductase specifically transfers H_R of the cofactor **59** to the substrate [95].

3.7
Isolation of Precorrin-8x and Proof of its Structure

Here again, *the remaining Cob enzymes were screened for their ability to use precorrin-6B* **63** *as substrate.* The successful assay was based on checking for transfer of the methyl group from [*methyl*-³H]SAM to precorrin-6B. It was likely that the next enzyme might be a methyltransferase because hydrogenobyrinic acid carries two methyl groups more than precorrin-6B. The selected enzyme proved to be encoded by the *cobL* gene and, when isolated on a preparative scale and purified to homogeneity, CobL smoothly transformed precorrin-6B **63** into a new substance carrying *two* more methyl groups. The new intermediate, named precorrin-8x, is efficiently converted enzymically into hydrogenobyrinic acid **60**, thus establishing its status as a new B_{12}-precursor [96, 97]. Dimethylation by one enzyme was not by now a surprise since precorrin-2 **12** is also biosynthesised in this way. However, the finding by mass spectrometry that precorrin-8x is a heptacarboxylic acid was certainly unexpected [97]. This means that the C-12 acetate had undergone decarboxylation and so CobL acts both as a methyltransferase and a decarboxylase, two very different processes. It may be that *cobL* has arisen by the fusion of two ancestral genes [96] and support for this interpretation comes from the description of two separate genes of the B_{12} pathway in *Salmonella typhimurium*, *cbiE* and *cbiT*, which show strong homology with the amino-terminal and carboxy-terminal regions of *cobL*, respectively [98].

The structural work on precorrin-8 x followed the ¹³C-labelling methods already fully described [99]. In this case, however, there were frustrating difficulties. The ester of precorrin-8 x was unstable so the NMR studies were carried out on the free acid. This isomerised in aqueous solution to generate first a mixture of at least five isomers and eventually one stable form. Separation of the former mixture by HPLC allowed each isomer to be tested as a biosynthetic precursor of hydrogenobyrinic acid **60** and it was found that the true precursor so identified could be stabilised at high pH for spectroscopic study. *One set of experiments differed from the earlier examples in that the labelled starting material was [2,3-$^{13}C_2$]ALA* **15b**, Scheme 24. The reason for using this particular labelling pattern was because information from ^{1}H-^{13}C correlation spectroscopy on the small scale involved is only available at and around ^{13}C-labelled sites (too many of the peaks overlap in the ^{1}H NMR spectrum for this to be used on its own). In this way, the nature of C-12 and its attached carbon in precorrin-8x could be studied [99] and these carbons are highlighted by using filled circles on structure **66b** in Scheme 24. In addition, a full ^{13}C-labelling study [99] on the stable form of precorrin-8x established its structure **67**. This knowledge of the structure of an isomer of the true intermediate was invaluable for the subsequent research on the latter.

The sum of all the results from ^{13}C-NMR and ^{1}H-NMR on both the true intermediate and its stable isomer led to structure **66** for precorrin-8x [99] and its key features are: (a) the two new methyl groups appear at C-5 and C-15, the latter being an sp^3 centre like C-10; (b) decarboxylation of the C-12 acetate group has been followed by tautomerisation to generate a methyl group at C-12; (c) there are several opportunities in structure **66** for double-bond tautomerism and also for epimerisation which probably account for the ready formation of the isomers reported above; (d) precorrin-8x **66** is isomeric with hydrogenobyrinic acid **60** and so is poised for the final rearrangement step that generates the corrin macrocycle. This step is covered in the following Sect. 3.8.

3.8
Migration of the C-11 Methyl Group to C-12 to Form the Corrin Macrocycle

Enzymic catalysis of a rearrangement reaction has special mechanistic fascination and, for vitamin B$_{12}$, a methyl group at C-11 of **66** shifts to C-12 of **60**, Scheme 25. This step clears the blockage to movement of the double bonds so tautomerisation can now occur, with thermodynamic gain, to form the extended conjugated system of hydrogenobyrinic acid **60**. *The enzyme responsible for*

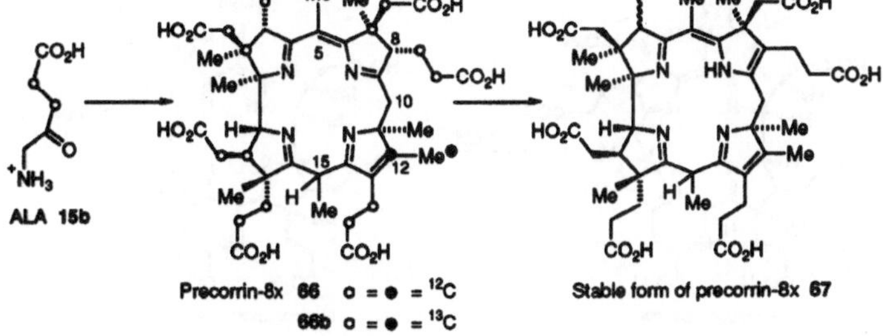

Scheme 24. Structure elucidation for precorrin-8×

catalysing this rearrangement process was eventually detected and its N-terminal sequence showed it to be encoded by the cobH gene [97]. Construction of a strain of Ps. denitrificans having cobH as the only amplified gene then allowed a protocol to be developed for isolation of pure CobH protein. This turned out [97] to be a relatively small monomer (M_r 22000 ± 1000) which further increases its attractiveness as a target for X-ray or NMR structure determination.

The mechanism for the migration of the methyl group from C-11 to C-12 is illustrated in Scheme 25 as an intramolecular [1,5]sigmatropic rearrangement. If this is true, the migration will be suprafacial and it was on this assumption that the C-11 methyl groups of precorrins-6A, -6B and -8x were assigned the α-configuration. It was essential to gain direct evidence that this was so because hard experience had taught the Paris-Cambridge teams that with vitamin B_{12}, one can assume nothing. Far too often, what had seemed likely proved to be wrong.

The decision between an intramolecular mechanism and an intermolecular one was made by a cross-over experiment. [$^{13}C_8$]Precorrin-3A was prepared enzymically [100] from [3-^{13}C]ALA and part was biosynthetically converted into [$^{13}C_8$]precorrin-6A using unlabelled SAM. The rest was similarly converted into [$^{13}C_8$,(CD$_3$)$_3$]precorrin-6A by using [methyl-$^2H_3$]SAM. A mixture of equal quantities of the two labelled samples of precorrin-6A was then enzymically converted into hydrogenobyrinic acid using unlabelled SAM. Mass spectrometric study of this final product proved that there had been no mixing of the CH_3 and CD_3 groups and hence that the methyl migration is intramolecular [101]. The illustrated [1,5]sigmatropic mechanism, Scheme 25, is thus supported and the α-configuration assigned to the C-11 methyl group of precorrin-6A 61 and later intermediates is secure.

It should be noted that the ^{13}C-labelling in the above studies was not, unlike earlier, for NMR analysis but simply to increase the mass so that the biosynthesised hydrogenobyrinic acid would be distinguished in the mass spectrum from the huge peak due to endogenous material appearing 8 mass units lower.

The position had now been reached where the biosynthetic route to hydrogenobyrinic acid 60, a precursor of vitamin B_{12}, had been mapped out from ALA through several steps to uro'gen III 10, before going forward along the "methylation pathway" to precorrin-2 12 and precorrin-3A 55. Then there was a gap to be filled before precorrin-6A 61 and subsequent intermediates. We will now

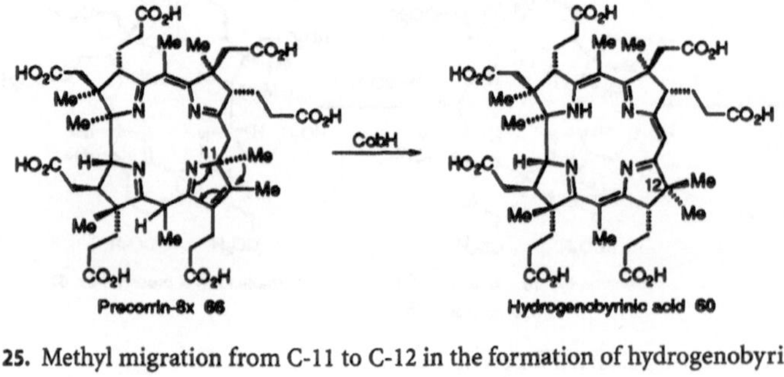

Scheme 25. Methyl migration from C-11 to C-12 in the formation of hydrogenobyrinic acid

describe how this final gap was filled to complete the entire biosynthetic pathway to coenzyme B$_{12}$.

3.9
Unexpected Structure of Precorrin-4

The approach leading to the discovery of precorrin-4, the tetramethylated B$_{12}$-intermediate, differed from those used previously. *A strain of Ps. denitrificans was constructed having the cobM gene deleted and thus unable to produce the CobM enzyme.* This strain yielded a cell-free enzyme system capable of converting precorrin-3A 55 into a new product, precorrin-4 [102]. Clearly, precorrin-4 is the substrate for CobM but is not further transformed, and so accumulates, because that enzyme is lacking. Precorrin-4 was readily oxidised by air to give its dehydro derivative, Factor IV, a step that could be reversed by enzymic reduction. Thus when Factor IV was incubated with the complete enzyme system, including NADH but lacking NADPH, precorrin-4 was regenerated and was carried forward to precorrin-6A 61. This conversion of precorrin-4 into an established intermediate for B$_{12}$ sets precorrin-4 securely on the biosynthetic pathway to the vitamin [102]. The relationship of precorrin-4 to Factor IV mirrors the relationships already well known for precorrin-2 12 and precorrin-3A 55 to their oxidised forms 56 and 57, Scheme 18; in these cases too the structural work had largely been carried out on the corresponding dehydro derivatives which also could be reduced enzymically to the original intermediates.

The structural studies [102] on Factor IV followed exactly the ^{13}C-labelling methods described already, now well developed, and led to the constitution 68, Scheme 26. This gave the crucial information that the formation of precorrin-4

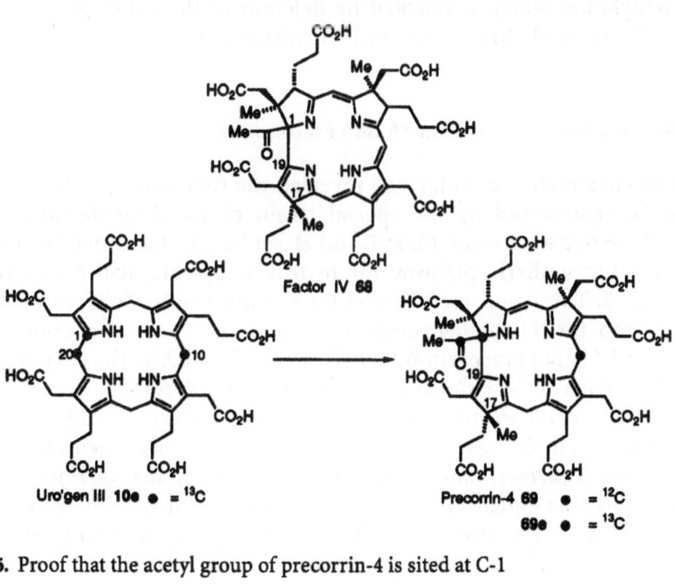

Scheme 26. Proof that the acetyl group of precorrin-4 is sited at C-1

involved an *oxidative* ring-contraction and C-17 methylation of precorrin-3A 55; thus structure 69 could confidently be deduced for precorrin-4 [103]. Then further NMR evidence in support of structure 69 was adduced [104]. The point was made earlier that an enzymic oxidation step must precede the formation of precorrin-6A 61; the structure of precorrin-4 69 now showed that this oxidation had already occurred by the tetramethylated stage. Besides this interest of structure 69, there were also two surprises, one being the unexpected early ring-contraction, the other being related to the acetyl group. Though formation of an acetyl residue during ring-contraction was expected on mechanistic grounds, for the NMR evidence to place it at C-1 was a surprise. Everyone's earlier speculations had placed the acetyl group at C-19 (actually in a postulated intermediate which would have been much later in the pathway). Interlocking evidence for the C-1 acetyl group came from the non-enzymic synthesis of $[1,10,20-^{13}C_3]$uro'-gen III 10e, which was converted enzymically via precorrin-3A into precorrin-4 69e, Scheme 26. The ^{13}C-NMR spectrum of the derived Factor IV showed doublets from the acetyl carbonyl group and from C-1, demonstrating direct bonding and thus making it certain that precorrin-4 69 carries its acetyl group at C-1 [105].

From the structure of precorrin-4 69 and its status as substrate for the CobM enzyme, clearly a methyltransferase from its sequence, the exact function of this enzyme could be deduced. In *Clostridium tetanomorphum* [106–108], *Pr. shermanii* [109] and *Ps. denitrificans* [88], the order of insertion of the various methyl groups into the corrin macrocycle, after the third one in precorrin-3A 55, had been revealed by pulse labelling and this order was the same in all three. The fourth one was placed at position 17, then at 12 α followed by 1 and, finally, 5 and 15. Precorrin-4 69 shows the C-17 methyl group in place and Sect. 3.5 reported that the 12 α-methyl of the corrin system is initially inserted at C-11. It is that C-11 methylation which is blocked by deletion of the *cobM* gene and hence CobM is the 11-methyltransferase of *Ps. denitrificans*.

3.10
Oxidative Conversion of Precorrin-3A into Precorrin-3B

The researches reviewed so far had revealed the functions of five of the eight enzymes overproduced by the special strain of *Ps. denitrificans* described earlier. These five were CobI, M, K, L and H, set here in the order in which they act on the biosynthetic pathway for hydrogenobyrinic acid 60 en route to vitamin B_{12} 3. Three enzymes were still available, CobG, CobF and CobJ, that together with the 11-methyltransferase, CobM, carry out the conversion of precorrin-3A 55 into precorrin-6A 61 via precorrin-4 69. So the enzymes for the two methylations at C-17 and at C-1 remained to be identified.

Strains of *Ps. denitrificans* were constructed in which *cobG*, *cobJ* and *cobF* had been separately amplified [110]. *The predicted amino acid sequences of the three corresponding enzymes showed that CobF and CobJ are methyltransferases whereas CobG was quite different.* Thus two methyltransferases were available for the C-17 and C-1 methylations with CobG being the remaining candidate to catalyse the oxidation step which precedes precorrin-4 69.

This analysis interlocked with the findings that (a) a strain of *Ps. denitrificans* from which *cobG* had been deleted accumulated precorrin-3A **55** [111] and (b) whereas CobJ alone had no effect on precorrin-3A [111], CobG and CobJ together converted it into precorrin-4 **69** and hence CobG acts before CobJ [110]. When tested, CobG alone converted precorrin-3A **55** into a new intermediate, named precorrin-3B, generated solely by oxidation without further methylation [112], observations which were independently confirmed [104]. Its molecular weight was 16 units higher than that of precorrin-3A corresponding to the addition of one oxygen atom to precorrin-3A **55**. The structure of precorrin-3B was determined by multiple ^{13}C-labelling as earlier but, in addition, a strong band at 1799 cm^{-1} in the IR-spectrum showed precorrin-3B to be a γ-lactone. The combined evidence established structure **70**, Scheme 27, for precorrin-3B [112, cf. 104]. Interestingly, the oxidation by CobG does not cause ring-contraction; rather the precorrin-3B molecule is primed ready for contraction in the next step, catalysed by CobJ, together with C-17 methylation to form precorrin-4 **69**.

The source of the oxygen atom added above to precorrin-3A was shown to be molecular oxygen [104] which was required to allow CobG to turn over. Also the oxygen atom was pinpointed as that in the C-20 hydroxyl group by using ^{18}O-isotope induced ^{13}C-NMR shifts [113]. Whether CobG acts as a mono-oxygenase or a dioxygenase was unknown as it was mechanistically possible that a second oxygen atom had been incorporated to provide the lactonic oxygen attached to C-1 of **70**. Because of the positioning of the ^{13}C-atoms in the foregoing NMR study, this would not have been detected. Conversion of precorrin-3A **55** into precorrin-3B **70** and precorrin-4 **69** with ^{18}O$_2$ followed by mass spectrometric study of both products showed that only one oxygen atom is incorporated [114]. This result together with that above showed that both lactonic oxygen atoms of precorrin-3B **70** are the original ones from the C-2 acetate side-chain of precorrin-3A **55**. These and similar studies also showed that the C-2 acetate oxygens are also retained in precorrin-4 **69** [114] and precorrin-5 [115], the topic of Sect. 3.11. The interest of these results and the following one will become clear from the discussion in Sect. 5.

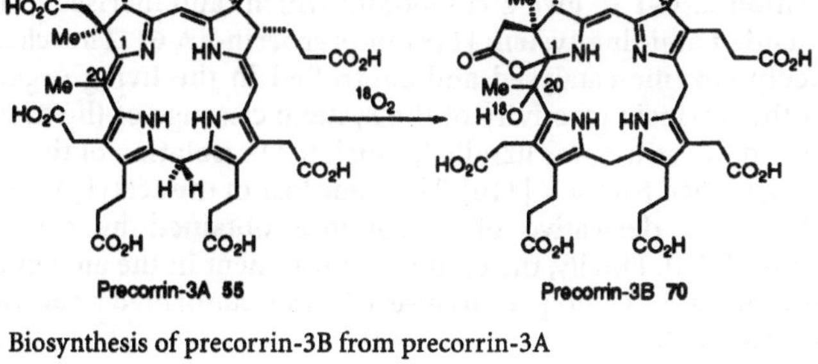

Scheme 27. Biosynthesis of precorrin-3B from precorrin-3A

Further experiments were designed to test whether the oxygens of the C-2 acetate in precorrin-3A 55 continue to be preserved beyond precorrin-5 through the remaining steps leading to hydrogenobyrinic acid 60, see Scheme 30. First, precorrin-3A 55 biosynthesised to carry eight $^{13}C^{18}O_2H$ groups, was converted into hydrogenobyrinic acid 60 by the overproduced enzymes of *Ps. denitrificans*. The ^{18}O-induced ^{13}C-NMR shifts for the seven distinguishable signals from the carboxyl groups of the labelled 60 showed that *both oxygen atoms* had been retained in all seven carboxyls, including the important acetate at C-2 [101]. This result was confirmed by mass measurements using electrospray mass spectrometry, an invaluable technique for handling such highly polar molecules.

The enzyme, CobG, central to the foregoing research, is a fascinating iron-sulfur protein with a brown-green colour. Its content of iron and sulfur points to one $[Fe_4S_4]$ or two $[Fe_2S_2]$ clusters being present [110] and no cofactors besides oxygen need be added for full activity.

3.11
Formation and Structure of Precorrin-5

By this stage, only the pentamethylated intermediate, precorrin-5, needed to be added to fill the last gap in the pathway to hydrogenobyrinic acid 60. As outlined in Sect. 3.9, conversion of precorrin-4 69 into precorrin-5 is blocked when CobM, the 11-methyltransferase, is excluded. Also, only one enzyme, the methyltransferase CobF, remains to catalyse the necessary C-1 methylation required to go forward from precorrin-5 to precorrin-6A 61. This knowledge led to two approaches for the production of precorrin-5, either from precorrin-3A 55 by incubation with CobG, CobJ, CobM and SAM but without CobF [110] or from precorrin-4 69 by using CobM and SAM [116]. The structure 71, Scheme 28, corresponding to 11-methyl-precorrin-4, was determined for precorrin-5 by the multiple ^{13}C-labelling approach as earlier [116].

Comparison of the structures of precorrin-5 71 and precorrin-6A 61 raises the problem of how the C-1 methyl group of the latter is introduced. An attractive rationalisation is shown in Scheme 28 initiated [117] by prototropic rearrangement of precorrin-5 71 to the isomer 72, now susceptible to hydrolytic elimination of the acetyl group as acetic acid by a reverse Claisen process. The extended enamine 73 so formed then allows [118] ready C-methylation at C-1 to give a compound which tautomerises to the more stable extended amidine system [119] in precorrin-6A 61. This chemistry is undoubtedly enzyme-catalysed and controlled in the living organism but draws on the intrinsic reactivity of the system: cleavage of the acetyl group occurred as precorrin-5 was handled, which led to isolation of the deacetylated product, named Factor V [110]. The same loss of the acetyl group was also observed with a derivative of precorrin-5 obtained by acid-catalysed esterification [116]. Finally, the eliminated fragment in the enzymic conversion of precorrin-5 71 into precorrin-6A 61 was confirmed as acetic acid by direct isolation [120].

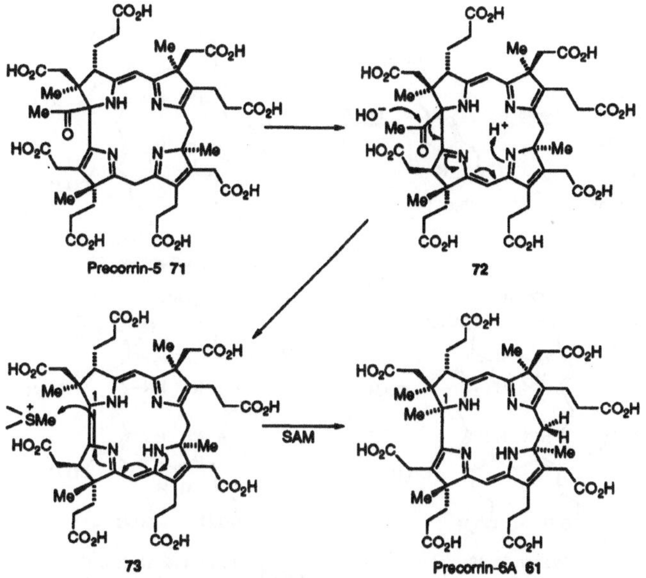

Scheme 28. Proposed mechanism for the introduction of the methyl group at C-1 of precorrin-6A

4
Complete Pathway from 5-Aminolaevulinic Acid to Hydrogenobyrinic Acid

All the B$_{12}$-intermediates discovered as a result of the experiments outlined in the main part of Sect. 3 have been assembled in Scheme 30 to show the entire biosynthetic pathway from precorrin-3A **55** through to hydrogenobyrinic acid **60**. By any measure, it is a remarkable sequence rich in surprises and unexpected chemistry. The early biosynthetic route from ALA **15** to precorrin-3A **55** is shown in Scheme 29, so Schemes 29 and 30 in combination show the complete story. Space only allows discussion of some of the more intriguing mechanistic aspects of Scheme 30; a fuller account is available [76].

The early steps, Scheme 29, used in common for haem, chlorophyll and vitamin B$_{12}$, involve condensation of two molecules of ALA **15** to form PBG **16** and four molecules of PBG are then combined to build uro'gen III **10** by way of hydroxymethylbilane **30**. Ring-closure of **30** to give **10** involves a fascinating intramolecular rearrangement. It is difficult to imagine a more compact route for the synthesis of uro'gen III **10**. This product stands at the branch point where the pathway to B$_{12}$ separates from that to haem and chlorophyll, the split being initiated by a double C-methylation of **10** by CobA to yield precorrin-2 **12**. A further C-methylation at C-20 by CobI generates precorrin-3A **55**. All three methylations, and also the remaining five needed to reach hydrogenobyrinic acid **60** are essentially C-methylations of enamines or extended enamines.

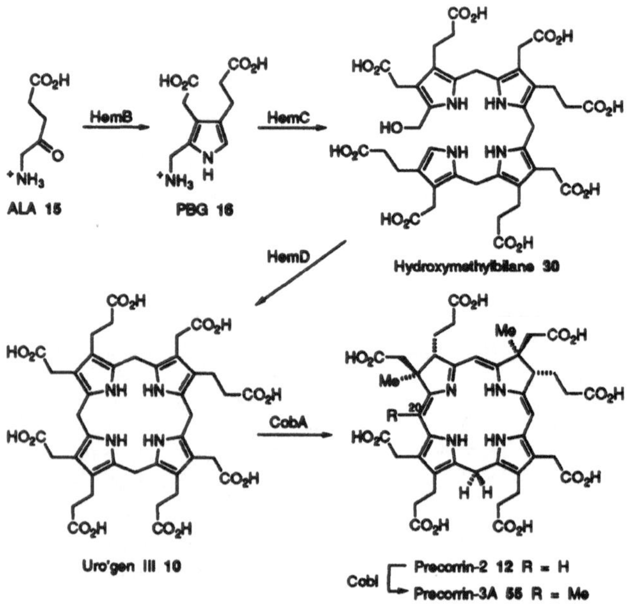

Scheme 29. Complete pathway from ALA through to precorrin-3A

Scheme 30 starts with the oxidation of precorrin-3A **55**, catalysed by CobG again with attack at the β-carbon of an enamine, to yield the lactonic precorrin-3B **70**; the possible involvement of lactones in B$_{12}$-biosynthesis was first suggested by Eschenmoser [119]. The oxidant is molecular oxygen and the single oxygen atom inserted forms the C-20 hydroxyl group. Precorrin-3B **70** is poised for a pinacol-type rearrangement [119] leading to early ring-contraction. This is illustrated in Scheme 30 as preceding the C-17 methylation that also occurs in this CobJ-catalysed generation of precorrin-4 **69**, which would allow the electron-rich pyrrole nucleus to be the migrating group. The next steps leading to precorrin-6A have already been discussed in Sect. 3.11, Scheme 28, and the reactions shown there share a common feature with all the chemistry in Scheme 30, in that each step sets up the reactivity needed for the next one. Thus, methylation of precorrin-4 **69** by CobM at C-11 gives precorrin-5 **71** shown in Scheme 28 to be ready for loss of the acetyl group so allowing C-1 methylation to afford precorrin-6A **61**. This was the intermediate whose discovery initiated the surge in knowledge of B$_{12}$-biosynthesis. Reduction of **61** by CobK, almost certainly as the C-18 protonated form, sets the oxidation level of the product, precorrin-6B **63**, to match that of hydrogenobyrinic acid **60**. The unusual enzyme, CobL, having both methyltransferase and decarboxylase activities, then acts on precorrin-6B **63** to insert methyl groups at C-5 and C-15 and to decarboxylate the C-12 acetate residue. It is attractive to suggest that the methylation at C-15 triggers the decarboxylation by converting ring-C into a protonated pyrrolenine ideally set up for decarboxylation. This chemistry leads to precorrin-8x **66** having ring-C as a

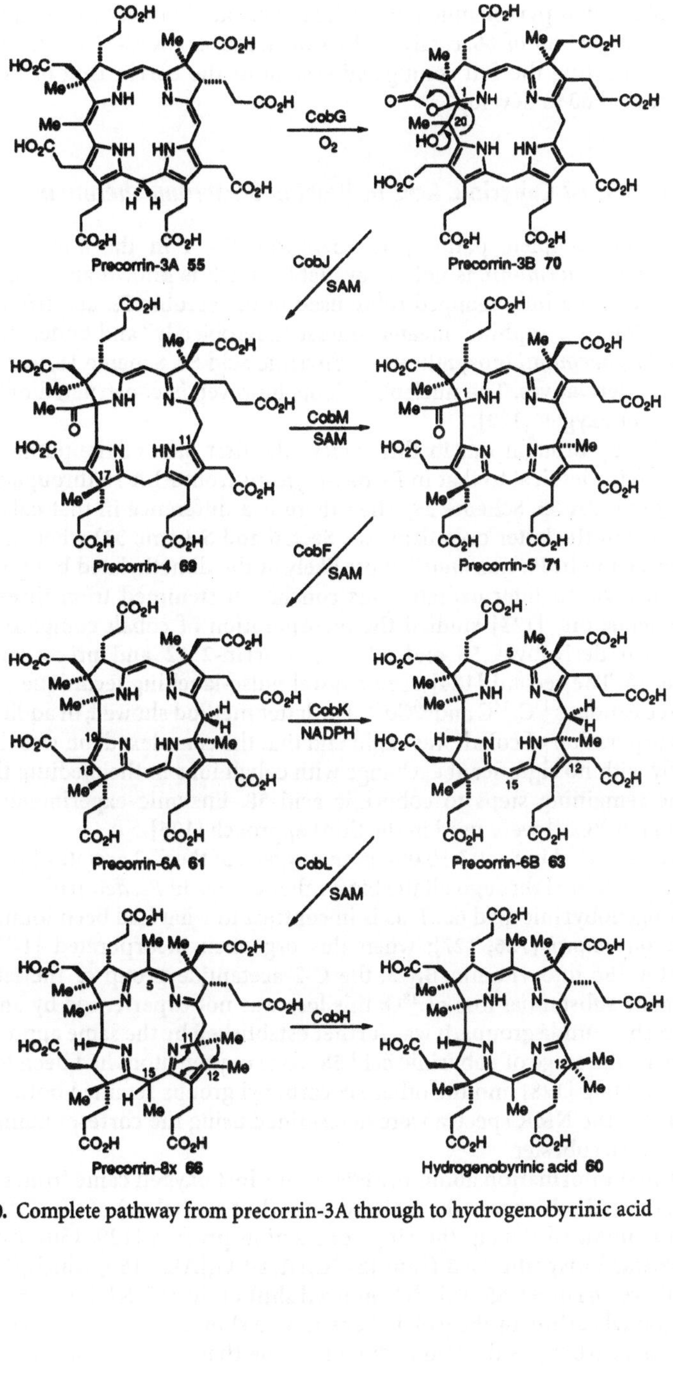

Scheme 30. Complete pathway from precorrin-3A through to hydrogenobyrinic acid

5,5-disubstituted pyrrolenine and such systems are known to rearrange [121]. The rearrangement of 66, catalysed by CobH, moves the C-11 methyl group to C-12, so allowing the full conjugated system of the corrin ring of hydrogen-obyrinic acid 60 to develop.

5
Biosynthesis of Cobyrinic Acid in *Propionibacterium shermanii*

Though the complete pathway to coenzyme B_{12} 4 in the micro-aerophilic organism, *Pr. shermanii*, is not yet available, much is known and a brief comparison with the fully mapped route used in the aerobic *Ps. denitrificans* now follows. "Micro-aerophilic" means "almost anaerobically" and under these conditions *Pr. shermanii* biosynthesises cobyrinic acid 58, Scheme 18, and some of its amide derivatives. The nucleotide loop, however, is constructed only in the presence of oxygen [122].

As already brought out in this review, the first part of the pathway in *Pr. shermanii* is identical to that in *Ps. denitrificans* from ALA 15 through six steps to precorrin-3A 55, Scheme 29. Then there is a difference in that cobalt is inserted late in the latter organism (see Sect. 6 and Scheme 33) whereas there is early insertion in *Pr. shermanii*, most likely at the dimethylated but possibly at the trimethylated intermediate. This conclusion stemmed from three sets of experiments; one [123] studied the incorporation of cobalt complexes of the aromatised derivatives 56 and 57 of precorrin-2 12 and precorrin-3A 55, Scheme 18. The second [124] used a novel pulse-labelling technique involving the three isotopes ^{13}C, ^{14}C and ^{60}Co^{2+}. The latter method showed, in addition, that the incorporation of cobalt was rapid and that the complexed ion was then held securely with no significant exchange with cobalt ions in the medium throughout the remaining steps to cobyrinic acid 58. Enzymic experiments on the insertion of ^{57}Co^{2+} were used in the third approach [125].

It was reported in Sect. 3.10 that both oxygens of the C-2 acetate of precorrin-3A 55 are retained through all the biosynthetic steps in *Ps. denitrificans* leading to hydrogenobyrinic acid 60. This is in contrast to what had been found earlier for *Pr. shermanii* [126, 127]; when this organism incorporated [1-^{13}C,1,1,4-^{18}O$_3$]ALA 15c into vitamin B_{12} 3, the C-2 acetamide group of the latter had undergone substantial loss of ^{18}O; this loss was not experienced by any of the other carboxamide groups. It was further established by the same approach that already by the stage of cobyrinic acid 58c, one oxygen atom had been lost from the C-2 acetate [128] and the other six carboxyl groups retained both oxygens, Scheme 31; the NMR spectra were determined using the corresponding hepta-methyl ester, cobester.

The first information about the fate of this lost oxygen came from studying the acetic acid, which was the isolated form (but see below) of the C_2 fragment that is eliminated during the ring-contraction process [129, 130]. Cobyrinic acid 58 was biosynthesised from [5-^{13}C,1,1,4-^{18}O$_3$]ALA 15d, which ^{13}C-labels C-20 of precorrin-3A 55. The ^{18}O-induced shift of the ^{13}C-NMR signal given by the carbonyl carbon of the isolated acetic acid showed that it had received one ^{18}O from a carboxyl side-chain which must be that of the C-2 acetate [131]. It

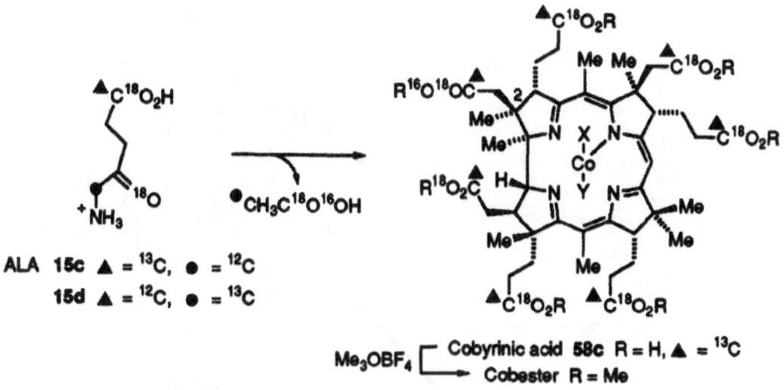

Scheme 31. Labelling with ¹⁸O and ¹³C to study the fate of the carboxyl oxygens

follows that the oxygen of this C-2 acetate becomes bonded to C-20 in the ring-contraction process rather than to C-1 as happens in *Ps. denitrificans*, see Sect. 3.10.

Evidence about the nature of this process has come from the isolation in a joint Stuttgart-Texas effort of a cobalto tetramethylated macrocycle from *Pr. shermanii*, shown by multiple ¹³C-labelling to have structure 74, Scheme 32 [132]. Presumably this compound, which is epimeric at C-8 relative to cobyrinic acid 58, is formed by enzymic methylation of 8-epi-precorrin-2 (8-epi 12) and the epimerisation caused derailment of the biosynthesis, allowing its isolation. The compound 74 is at the dehydro level of oxidation relative to what is needed for conversion forward into cobyrinic acid 58 and is considered to have been oxidised during isolation just as, for example, precorrin-2 12 was in the early studies. Its conversion into cobyrinic acid 58 thus requires both enzymic reduction and epimerisation at C-8, which would explain the low observed level of incorporation (0.6%).

A mechanism for the formation of 74 has been proposed [132] which has much in common with earlier ideas based on model studies by Eschenmoser and coworkers [119, 133, 134]. In addition, it is suggested that the dihydro form of 74 undergoes methylation at C-11, similar to that in *Ps. denitrificans* to yield 75. This could undergo a retro-aldol step to release acetaldehyde, Scheme 32. Indeed, when the 2-carbon fragment was immediately trapped, it *was* found to be acetaldehyde [135]; the earlier isolation of acetic acid resulted from oxidation of acetaldehyde by long incubation with the mixture of enzymes.

The outcome of the retro-aldol step would be the compound 76 in Scheme 32, which is the cobalt complex of 73, ready for C-1 methylation (cf. Scheme 28) to generate cobalto-precorrin-6A 77. If this is correct, the pathways for biosynthetic modification of the organic macrocycles in the two organisms merge again at 76.

This remerging of the chemistry in the two organisms had been supported by earlier studies [125] in which some of the intermediates isolated from *Ps. denitrificans* were incorporated into cobyrinic acid 58 by an enzyme preparation from *Pr. shermanii*. Precorrin-6A 61, precorrin-6B 63 and precorrin-8x 66,

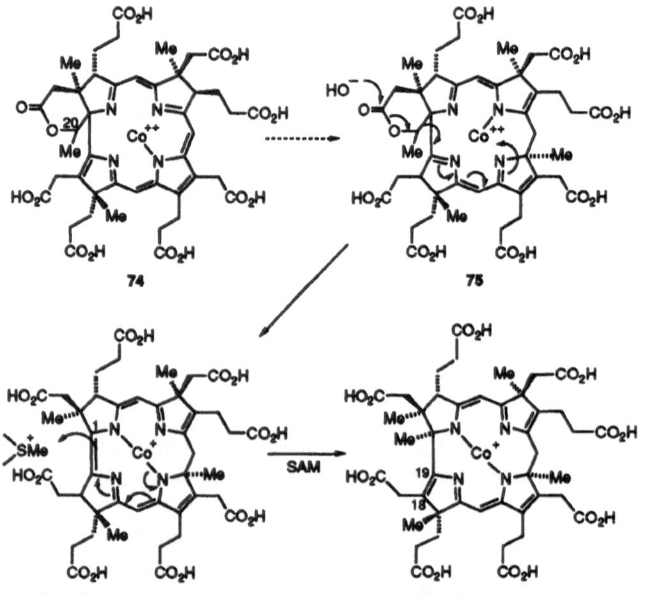

Scheme 32. Proposed mechanism for the conversion of 74 into cobalto-precorrin-6A

Scheme 30, in $^3H/^{14}C$ doubly labelled form, were all incorporated into cobyrinic acid 58 with unchanged $^3H:^{14}C$ ratio. The incorporation levels (0.9–2.7%) were reasonable bearing in mind that these three macrocycles need to sequester cobalt from the medium to enter the biosynthetic pathway in *Pr. shermanii*.

In addition, it was proved [136] that an enzymic reduction which transfers a hydride equivalent to C-19 occurs as cobyrinic acid 58 is biosynthesised in *Pr. shermanii*. This finding fits perfectly with conversion of cobalto-precorrin-6A 77 into cobalto-precorrin-6B (Co-63); the equivalent reductive step in *Ps. denitrificans* was discussed in Sect. 3.6. Also, the stereospecificity of the transfer, employing 4-H_R of the reduced cofactor, was the same in both biological systems [136].

Thus, the picture at present is that the differences between *Ps. denitrificans* and *Pr. shermanii* are (a) cobalt is inserted late in the former and early in the latter and (b) the ring-contraction process involves intriguingly different chemistry around C-1, C-20 and the C-2 acetate side-chain in the two organisms before the same organic macrocycles are used again. Of the 22 or so steps from ALA 15 to coenzyme B_{12} 4, it already appears that some 16 involve either identical substances in the two organisms or they use the same organic ligand. The biosynthesis of coenzyme B_{12} in the two organisms is thus based on the same biosynthetic theme with two variations. This reinforces our thinking that the pathways used by the two organisms are quite close in evolutionary terms.

6
Biosynthesis of Coenzyme B$_{12}$ from Hydrogenobyrinic Acid

Our plan was to focus this Chapter largely on the biosynthesis of hydrogeno-byrinic acid **60**. Nevertheless, it is important to show even briefly how coenzyme B$_{12}$ **4**, Scheme 33, is constructed from **60**. Besides the intrinsic value of this knowledge, it considerably helped the researches described in Sect. 3 because *by detecting those genes acting beyond hydrogenobyrinic acid **60**, the problem of pin-pointing the ones required to build the corrin macrocycle **60** was greatly simplified.*

Scheme 33 shows the steps elucidated [76] for the biosynthesis of coenzyme B$_{12}$ **4** from hydrogenobyrinic acid **60** in *Ps. denitrificans*. This collects together the knowledge generated for this organism by the French teams of Blanche and Crouzet. A more complete review of this work and that of others is available [76].

The substrate for cobalt insertion was found to be the *a,c*-diamide **78** gener-ated from hydrogenobyrinic acid **60** by CobB. Remarkably, the *superficially simple* step of cobalt insertion requires the cooperation of three proteins, CobN, CobS and CobT, to form a cobalt cheletase system which is ATP-dependent. The product, cob(II)yrinic acid *a,c*-diamide, is reduced to the Co(I) state by a flavin dependent enzyme (gene as yet unknown), which was isolated as pure protein. The highly nucleophilic Co(I)-complex was thus ready for adenosylation at

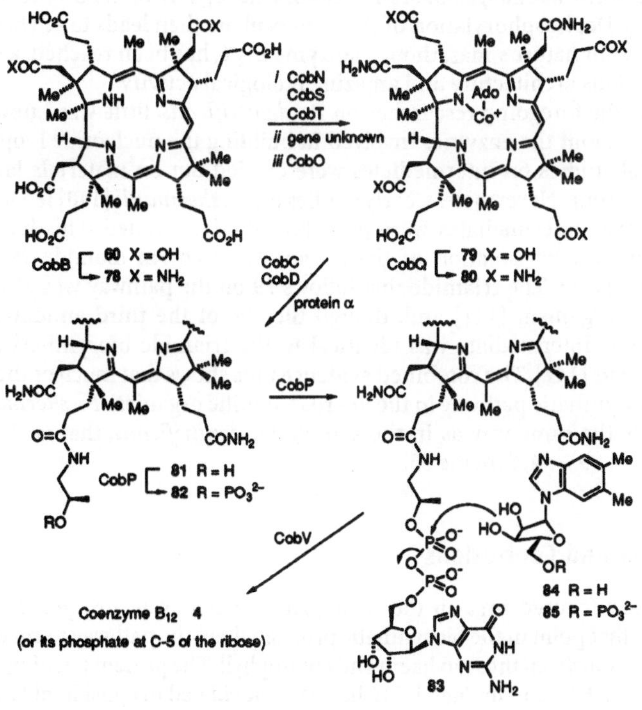

Scheme 33. Complete pathway from hydrogenobyrinic acid to coenzyme B$_{12}$

cobalt by ATP catalysed by CobO to afford adenosylcobyrinic acid *a,c*-diamide
79. Amidation of three of the propionate side-chains and the one remaining
acetate side-chain is then carried out by CobQ to form adenosylcobyric acid **80**
leaving the fourth propionate, on ring D, available for attachment of the propanolamine unit. The latter is derived from threonine but, since no threonine
decarboxylase has ever been found, a 3-step mechanism was considered involving enzymic dehydrogenation of threonine to 2-amino-3-ketobutyrate then
decarboxylation followed by reduction to aminopropanol. Direct studies on *Pr.*
shermanii [137] and *Salmonella typhimurium* [138] showed that this sequence
is not followed, however. Fortunately, relief from the impasse has appeared in a
recent report [139] that *O*-phospho-L-threonine is involved in the process. The
attachment of the propanolamine unit requires a protein designated α together
with CobC and CobD to afford adenosylcobinamide **81**. Then follows the ATP-
dependent phosphorylation of the propanolamine hydroxyl group catalysed by
CobP to yield **82**. CobP is yet another multifunctional Cob enzyme in that it is
also responsible for the transfer of a guanosyl monophosphate (GMP) unit from
the triphosphate (GTP) to **82** affording adenosyl-GDP-cobinamide **83**.

A separate pathway has evolved to build α-ribazole **84**, unusual both in its
base, 5,6-dimethylbenzimidazole, and in the α-linkage to ribose. Either α-
ribazole **84** or its 5′-phosphate **85** is able to accept the adenosylcobinamide
phosphate residue from adenosyl-GDP-cobinamide **83**, a transfer catalysed by
CobV, to form adenosylcobalamin (coenzyme B_{12}, **4**) or its 5′-phosphate, respectively. Dephosphorylation of the 5′-phosphate then leads to **4**. Thus the end
product from nature's marathon, coenzyme B_{12} **4**, has been reached, a structure
of marvellous architecture and amazing biological activity.

Before the foregoing researches on *Ps. denitrificans*, little was known in any
organism about the enzymes involved in building the nucleotide loop of **4** and
structural studies on intermediates were carried out on materials lacking the
adenosyl group. Nevertheless, early studies on *Pr. shermanii* [140] led to the view
that the true intermediates were probably all adenosylated. This has received
direct support by isolation of the adenosylated corrin diamide **79** from *Pr.*
shermanii [141]. The triamide that follows **79** on the pathway was also isolated
from this organism [141] and, though the site of the third amidation is unknown, this intermediate was identical to the triamide biosynthesised by *Ps.*
denitrificans [142]. The combined evidence thus shows that much of the late part
of the biosynthetic pathway in the microaerophilic organism *Pr. shermanii* flows
in exactly the same way as in the aerobic *Ps. denitrificans*, that is **79** → **80** →
81 → **82** → **83** → **4**, Scheme 33.

7
Summary and Conclusions

Our review focused first on the tetrapyrrolic macrocycle, uro'gen III **10** which
stands at that point in the biosynthetic process where the pathway to coenzyme B_{12}
branches away from those to haem and chlorophyll. The present position is that the
route from ALA **15** to uro'gen III **10** has been elucidated in considerable detail and
substantial knowledge has been gained about the mechanisms of the processes

involved. The enzymes that catalyse these reactions are now available in pure state and, with the identification of their genes, have been overproduced, so opening the door to crystallisation of the protein and X-ray structure analysis. There will surely be further progress in structural work on these enzymes and one particularly looks forward to the day when the active site of uro'gen III synthase can be seen. Further, the Laue X-ray approach for the study of at least one of these enzymes in action as it binds and manipulates the substrate offers exciting opportunities.

Moving forward from uro'gen III, this review showed that there has been a transformation in knowledge of the biosynthesis of vitamin B_{12} 3 and coenzyme B_{12} 4; progress has been faster and more complete than anyone in the field dared to imagine.

In *Ps. denitrificans*, the aerobic organism used for most of these recent researches, the conversion of ALA 15 through to coenzyme B_{12} 4 involves 22 steps and this entire pathway has been fully elucidated. The enzymes that catalyse essentially every step have been identified together with the corresponding genes. In the majority of cases, the genes have been overexpressed and the enzymes isolated, purified and characterised. Complementary to this biological research has been the chemical effort which has led to the elucidation of the structures of all the intermediates on the pathway and, starting with uro'gen III, there are 19 main intermediates leading to coenzyme B_{12} 4. Also, the mechanism and stereochemistry of a number of the important reactions involved in producing and transforming these intermediates have been worked out. However, the impression must not be given that the biological and chemical researches were in any way separate; they were tightly interlocked and each was essential for the other. Indeed, it was the synergistic combination of genetics, molecular biology, enzymology, synthetic and structural chemistry, isotopic labelling and NMR spectroscopy that led to the dramatic progress in solving the problem of B_{12}-biosynthesis in *Ps. denitrificans*.

Much is also known about the route from ALA to coenzyme B_{12} in the microaerophilic organism *Pr. shermanii*. Of those steps that have been studied so far, a substantial majority occur using either identical intermediates to those established in *Ps. denitrificans* or they involve the cobalt complexes of the same organic ligands. There are differences, however, in that cobalt is inserted far earlier on the pathway in *Pr. shermanii* and an interesting modified mechanism is used for the ring contraction in this micro-aerophilic system.

The final message is that the researches reviewed demonstrate the enormous power of the tools that are now available to biologists and bio-organic chemists for discovering how even the most complex substances are built by living systems.

8
References

1. Leeper FJ (1989) Nat Prod Rep 6:171; (1987) 4:441;(1985) 2:561 and 19
2. Jordan PM (ed) (1991) Biosynthesis of tetrapyrroles. Elsevier, Amsterdam
3. Leeper FJ (1991) Intermediate steps in the biosynthesis of chlorophylls. In: Scheer H (ed) Chlorophylls. CRC Press, Boca Raton, p 407
4. Battersby AR, McDonald E (1975) Biosynthesis of porphyrins, chlorins and corrins. In: Smith KM (ed) Porphyrins and metalloporphyrins. Elsevier, Amsterdam, p 61

5. (1994) The biosynthesis of the tetrapyrrole pigments. Ciba Foundation Symposium, vol 180. Wiley, Chichester
6. Battersby AR, McDonald E (1982) Biosynthesis of the corrin macrocycle. In: Dolphin D (ed) B_{12}, vol 1. Wiley, London, p 107
7. Spencer JB, Stolowich NJ, Roessner CA, Scott AI (1993) FEBS Lett 335:57
8. Woodcock SC, Warren MJ (1996) Biochem J 313:415
9. Yap-Bondoc F, Bondoc LL, Timkovich R, Baker DC, Hebbler A (1990) J Biol Chem 265:13498
10. Micklefield J, Beckmann M, Mackman RL, Block MH, Leeper FJ, Battersby AR (1997) J Chem Soc, Perkin Trans 1:2123
11. Pfaltz A, Kobelt A, Hüster R, Thauer RK (1987) Eur J Biochem 170:459
12. Stupperich E, Eisinger HJ, Schurr S (1990) FEMS Microbiol Rev 87:355
13. Kannangara CG, Andersen RV, Pontoppidan B, Willows R, von Wettstein D (1994) Enzymic and mechanistic studies on the conversion of glutamate to 5-aminolaevulinate. In: The biosynthesis of the tetrapyrrole pigments. Ciba Foundation Symposium, vol 180. Wiley, Chichester, p 3
14. Mau YHL, Wang WY (1988) Plant Physiol 86:793
15. Bull AD, Breu V, Kannangara CG, Rogers LJ, Smith AJ (1990) Arch Microbiol 154:56
16. Spencer P, Jordan PM (1995) Biochem J 305:151
17. Jaffe EK, Ali S, Mitchell LW, Taylor KM, Volin M, Markham GD (1995) Biochemistry 34:244
18. Appleton D, Leeper FJ (1996) Chem Commun 303
19. Appleton D, Leeper FJ (1996) Bioorg Med Chem Lett 6:1191
20. Nandi DL, Shemin D (1968) J Biol Chem 243:1236
21. Jordan PM, Seehra JS (1980) FEBS Lett 114:283
22. Jordan PM, Seehra JS (1980) J Chem Soc, Chem Commun 240
23. Dent AJ, Beyersmann D, Block C, Hasnain SS (1990) Biochemistry 29:7822
24. Seehra JS, Jordan PM (1981) Eur J Biochem 113:435
25. Nayar P, Stolowich NJ, Scott AI (1995) Bioorg Med Chem Lett 5:2105
26. Carrell HL, Glusker JP, Shimoni L, Keefe LJ, Afshar C, Volin M, Jaffe EK (1996) Acta Cryst D, Biol Crystallog 52:419
27. Senior NM, Brocklehurst K, Cooper JB, Wood SP, Erskine P, Shoolingin Jordan PM, Thomas PG, Warren MJ (1996) Biochem. J. 320:401
28. Battersby AR, Leeper FJ (1990) Chem Rev 90:1261
29. (1983) Biochem J 209:II
30. Hart GJ, Miller AD, Leeper FJ, Battersby AR (1987) J Chem Soc, Chem Commun 1762
31. Hart GJ, Miller AD, Beifuss U, Leeper FJ, Battersby AR (1990) J Chem Soc, Perkin Trans 1 1979
32. Miller AD, Leeper FJ, Battersby AR (1989) J Chem Soc, Perkin Trans 1 1943
33. Jordan PM, Warren MJ (1987) FEBS Lett 225:87
34. Hart GJ, Miller AD, Battersby AR (1988) Biochem J 252:909
35. Beifuss U, Hart GJ, Miller AD, Battersby AR (1988) Tetrahedron Lett 29:2591
36. Jordan PM, Warren MJ, Williams HJ, Stolowich NJ, Roessner CA, Grant SK, Scott AI (1988) FEBS Lett 235:189
37. Scott AI, Stolowich NJ, Williams HJ, Gonzalez MD, Roessner CA, Grant SK, Pichon C (1988) J Am Chem Soc 110:5898
38. Miller AD, Hart GJ, Packman LC, Battersby AR (1988) Biochem J 254:915
39. Scott AI, Roessner CA, Stolowich NJ, Karuso P, Williams HJ, Grant SK, Gonzalez MD, Hoshino T (1988) Biochemistry 27:7984
40. Shoolingin Jordan PM, Warren MJ, Awan SJ (1996) Biochem J 316:373
41. Louie G, Brownlie PD, Lambert R, Cooper JB, Blundell TL, Wood SP, Warren MJ, Woodcock SC, Jordan PM (1992) Nature 359:33
42. Hädener A, Matzinger PK, Malashkevich VN, Louie GV, Wood SP, Oliver P, Alefounder PR, Pitt AR, Abell C, Battersby AR (1993) Eur J Biochem 211:615
43. Louie GV, Brownlie PD, Lambert R, Cooper JB, Blundell TL, Wood SP, Malashkevich VN, Hädener A, Warren MJ, Shoolingin Jordan PM (1996) Proteins - Struct Funct Genet 25:48

44. Hädener A, Matzinger P, Battersby AR, McSweeney S, Thompson A, Harrop S, Cassetta A, Deacon A, Hunter W, Peterson M, Helliwell J (1997) Acta Crystallog. D, submitted
45. Lambert R, Brownlie PD, Woodcock SC, Louie GV, Cooper JC, Warren MJ, Jordan PM, Blundell TL, Wood SP (1994) Structural studies on porphobilinogen deaminase. In: The biosynthesis of the tetrapyrrole pigments. Ciba Foundation Symposium, vol 180. Wiley, Chichester, p 97
46. Jordan PM, Woodcock SC (1991) Biochem J 280:445
47. Lander M, Pitt AR, Alefounder PR, Bardy D, Abell C, Battersby AR (1991) Biochem J 275:447
48. Woodcock SC, Jordan PM (1994) Biochemistry 33:2688
49. Miller AD, Packman LC, Hart GJ, Alefounder PR, Abell C, Battersby AR (1989) Biochem J 262:119
50. Hädener A, Alefounder PR, Hart GJ, Abell C, Battersby AR (1990) Biochem J 271:487
51. Leeper FJ (1996) unpublished results
52. Evans JNS, Fagerness PE, Mackenzie NE, Scott AI (1985) Magn Reson Chem 23:939
53. Battersby AR (1986) Ann N Y Acad Sci 471:138
54. Warren MJ, Gul S, Aplin RT, Scott AI, Roessner CA, O'Grady P, Shoolingin Jordan PM (1995) Biochemistry 34:11288
55. Schauder J-R, Jendrezejewski S, Abell A, Hart GJ, Battersby AR (1987) J Chem Soc, Chem Commun 436
56. Nieh YP, Helliwell JR, Cassetta A, Raftery J, Hädener A, Niemann AC, Battersby AR, Carr PD, Wulff M, Ursby T, Moy JP, Thompson AW, in preparation
57. Leeper FJ, Rock M, Appleton D (1996) J Chem Soc, Perkin Trans 1 2633
58. Warren MJ, Jordan PM (1988) Biochemistry 27:9020
59. Wang JJ, Scott AI (1994) Tetrahedron 50:6181
60. Battersby AR, Fookes CJR, Matcham GWJ, McDonald E, Gustafson-Potter KE (1979) J. Chem. Soc., Chem. Commun. 316
61. Leeper FJ, Rock M (1996) J Chem Soc, Perkin Trans 1 2643
62. Bardy D, Pitt AR, Leeper FJ, Battersby AR (1991) unpublished results, Cambridge
63. Clemens KR, Pichon C, Jacobson AR, Yonhin P, Gonzalez MD, Scott AI (1994) Bioorg Med Chem Lett 4:521
64. Battersby AR, Fookes CJR, Matcham GWJ, Pandey PS (1981) Angew Chem, Int Ed Engl 20:293
65. Battersby AR, Fookes CJR, Pandey PS (1983) Tetrahedron 39:1919
66. Pichon C, Atshaves BP, Xue TH, Stolowich NJ, Scott AI (1994) Bioorg Med Chem Lett 4:1105
67. Mathewson JH, Corwin AH (1961) J Am Chem Soc 83:135
68. Stark WM, Baker MG, Leeper FJ, Raithby PR, Battersby AR (1988) J Chem Soc, Perkin Trans 1 1187
69. Stark WM, Hawker CJ, Hart GJ, Philippides A, Petersen PM, Lewis JD, Leeper FJ, Battersby AR (1993) J Chem Soc, Perkin Trans 1 2875
70. Cassidy MA, Crockett N, Leeper FJ, Battersby AR (1996) J Chem Soc, Perkin Trans 1 2079
71. Spivey AC, Capretta A, Frampton CS, Leeper FJ, Battersby AR (1996) J Chem Soc, Perkin Trans 1 2091
72. See for example: Xu WM, Astrin KH, Desnick RJ (1996) Human Mutation 7:187
73. Leeper FJ (1994) Evidence for a spirocyclic intermediate in the formation of uro'gen III by cosynthetase. In:The biosynthesis of the tetrapyrrole pigments. Ciba Foundation Symposium, vol 180. Wiley, Chichester, p 111
74. Tietze LF, Geissler H, Schulz G (1994) Pure Appl Chem 66:2303
75. Stamford NPJ, Capretta A, Battersby AR (1995) Eur J Biochem 231:236
76. Blanche F, Cameron B, Crouzet J, Debussche L, Thibaut D, Vuilhorgne M, Leeper FJ, Battersby AR (1995) Angew Chem, Int Ed Engl 35:383
77. Battersby AR (1986) Acc Chem Res 19:147
78. Scott AI (1990) Acc Chem Res 23:308
79. Cameron B, Briggs K, Pridmore S, Brefort G, Crouzet J (1989) J Bacteriol 171:547

80. Crouzet J, Cameron B, Cauchois L, Rigault S, Rouyez M-C, Blanche F, Thibaut D, Debussche L (1990) J Bacteriol 172:5980
81. Crouzet J, Cauchois L, Blanche F, Debussche L, Thibaut D, Rouyez M-C, Rigault S, Mayaux J-F, Cameron B (1990) J Bacteriol 172:5968
82. Crouzet J, Levy-Schil S, Cameron B, Cauchois L, Rigault S, Rouyez M-C, Blanche F, Debussche L, Thibaut D (1991) J Bacteriol 173:6074
83. Cameron B, Blanche F, Rouyez M-C, Bisch D, Famechon A, Couder M, Cauchois L, Thibaut D, Debussche L, Crouzet J (1991) J Bacteriol 173:6066
84. Cameron B, Guilholt C, Blanche F, Cauchois L, Rouyez M-C, Rigault S, Levy-Schil S, Crouzet J (1991) J Bacteriol 173:6058
85. Blanche F, Debussche L, Thibaut D, Crouzet J, Cameron B (1989) J Bacteriol 171:4222
86. Thibaut D, Couder M, Crouzet J, Debussche L, Cameron B, Blanche F (1990) J Bacteriol 172:6245
87. Weaver GW, Blanche F, Thibaut D, Debussche L, Leeper FJ, Battersby AR (1990) J Chem Soc, Chem Commun 1125–27
88. Blanche F, Thibaut D, Fréchet D, Vuilhorgne M, Crouzet J, Cameron B, Hlineny K, Traub-Eberhard U, Zboron M, Müller G (1990) Angew Chem, Int Ed Engl 29:884
89. Thibaut D, Debussche L, Blanche F (1990) Proc Natl Acad Sci USA 87:8795
90. Thibaut D, Blanche F, Debussche L, Leeper FJ, Battersby AR (1990) Proc Natl Acad Sci USA 87:8800
91. Blanche F, Kodera M, Couder M, Leeper FJ, Thibaut D, Battersby AR (1992) J Chem Soc, Chem Commun 138
92. Blanche F, Thibaut D, Famechon A, Debussche L, Cameron B, Crouzet J (1992) J Bacteriol 174:1036
93. Thibaut D, Kiuchi F, Debussche L, Leeper FJ, Blanche F, Battersby AR (1992) J Chem Soc, Chem Commun 139
94. Weaver GW, Leeper FJ, Battersby AR, Blanche F, Thibaut D, Debussche L (1991) J Chem Soc, Chem Commun 976
95. Kiuchi F, Thibaut D, Debussche L, Leeper FJ, Blanche F, Battersby AR (1992) J Chem Soc, Chem Commun 306
96. Blanche F, Famechon A, Thibaut D, Debussche L, Cameron B, Crouzet J (1992) J Bacteriol 174:1050
97. Thibaut D, Couder M, Famechon A, Debussche L, Cameron B, Crouzet J, Blanche F (1992) J Bacteriol 174:1043
98. Roth JR, Lawrence JG, Rubenfield M, Kieffer-Higgins S, Church GM (1993) J Bacteriol 175:3303
99. Thibaut D, Kiuchi F, Debussche L, Blanche F, Kodera M, Leeper FJ, Battersby AR (1992) J Chem Soc, Chem Commun 982
100. Stamford NPJ, Crouzet J, Cameron B, Alanine AID, Pitt AR, Yeliseev AA, Battersby AR (1996) Biochem J 313:335
101. Li Y, Alanine AID, Balachandran S, Vishwakarma RA, Leeper FJ, Battersby AR (1994) J Chem Soc, Chem Commun 2507
102. Thibaut D, Debussche L, Fréchet D, Herman F, Vuilhorgne M, Blanche F (1993) J Chem Soc, Chem Commun 513
103. Battersby AR (1994) New intermediates in the B$_{12}$ pathway. In: The biosynthesis of the tetrapyrrole pigments. Ciba Foundation Symposium, vol 180. Wiley, Chichester, p 267
104. Scott AI, Roessner CA, Stolowich NJ, Spencer JB, Min C, Ozaki SI (1993) FEBS Lett 331:105
105. Alanine AID, Ichinose K, Thibaut D, Debussche L, Stamford NPJ, Leeper FJ, Blanche F, Battersby AR (1994) J Chem Soc, Chem Commun 193
106. Uzar HC, Battersby AR (1982) J Chem Soc, Chem Commun 1204
107. Uzar HC, Battersby AR (1985) J Chem Soc, Chem Commun 585
108. Uzar HC, Battersby AR, Carpenter TA, Leeper FJ (1987) J Chem Soc, Perkin Trans 1 1689
109. Scott AI, Mackenzie NE, Santander PJ, Fagerness PE, Müller G, Schneider E, Sedlmeier R, Wörner G (1984) Bioorg Chem 12:356

110. Debussche L, Thibaut D, Cameron B, Crouzet J, Blanche F (1993) J Bacteriol 175:7430
111. Blanche F, Crouzet J, Cameron B, Thibaut D, unpublished results
112. Debussche L, Thibaut D, Danzer M, Debu F, Fréchet D, Herman F, Blanche F, Vuilhorgne M (1993) J Chem Soc, Chem Commun 1100
113. Spencer JB, Stolowich NJ, Roessner CA, Min CH, Scott AI (1993) J Am Chem Soc 115:11|610
114. Alanine AID, Li Y, Stamford NPJ, Leeper FJ, Blanche F, Debussche L, Battersby AR (1994) J Chem Soc, Chem Commun 1649
115. Spencer JB, Stolowich NJ, Santander PJ, Pichon C, Kajiwara M, Tokiwa S, Takatori K, Scott AI (1994) J Am Chem Soc 116:4991
116. Min C, Atshaves BP, Roessner CA, Stolowich NJ, Spencer JB, Scott AI (1993) J Am Chem Soc 115:10|380
117. Arigoni D (1994) In: The biosynthesis of the tetrapyrrole pigments. Ciba Foundation Symposium, vol 180. Wiley, Chichester, p 307
118. Battersby AR (1994) In: The biosynthesis of the tetrapyrrole pigments. Ciba Foundation Symposium, vol 180. Wiley, Chichester, p 307
119. Eschenmoser A (1988) Angew Chem, Int Ed Engl 27:5
120. Li Y, Stamford NPJ, Battersby AR (1995) J Chem Soc, Perkin Trans 1:283
121. Battersby AR, Baker MG, Broadbent HA, Fookes CJR, Leeper FJ (1987) J Chem Soc, Perkin Trans 1:2027
122. Perlman D (1971) Methods Enzymol 18C:75
123. Müller G, Zipfel F, Hlineny K, Savvidis E, Hurtle R, Traub-Eberhard U, Scott AI, Williams HJ, Stolowich NJ, Santander PJ, Warren MJ, Blanche F, Thibaut D (1991) J Am Chem Soc 113:9893
124. Balachandran S, Vishwakarma RA, Monaghan SM, Prelle A, Stamford NPJ, Leeper FJ, Battersby AR (1994) J Chem Soc, Perkin Trans 1 487
125. Blanche F, Thibaut D, Debussche L, Hertle R, Zipfel F, Müller G (1993) Angew Chem, Int Ed Engl 32:1651
126. Kurumaya K, Okazaki T, Kajiwara M (1989) Chem Pharm Bull Jpn 37:1151
127. Scott AI, Stolowich NJ, Atshaves BP, Karuso P, Warren MJ, Williams HJ, Kajiwara M, Kurumaya K, Okazaki T (1991) J Am Chem Soc 113:9891
128. Vishwakarma RA, Balachandran S, Alanine AID, Stamford NPJ, Kiuchi F, Leeper FJ, Battersby AR (1993) J Chem Soc, Perkin Trans 1 2893
129. Mombelli L, Nussbaumer C, Weber H, Müller G, Arigoni D (1981) Proc Natl Acad Sci USA 78:9
130. Battersby AR, Bushell MJ, Jones C, Lewis NG, Pfenninger A (1981) Proc Natl Acad Sci USA 78:13
131. Broers S, Berry A, Arigoni D (1994) In: The biosynthesis of the tetrapyrrole pigments. Ciba Foundation Symposium, vol 180. Wiley, Chichester, p 280
132. Scott AI, Stolowich NJ, Wang JJ, Gawatz O, Fridrich E, Müller G (1996) Proc Natl Acad Sci USA 93:14316
133. Yamada Y, Miljkovic D, Wehrli P, Golding B, Löliger P, Keese R, Müller K, Eschenmoser A (1969) Angew Chem, Int Ed Engl 8:343
134. Ofner S, Rasetti V, Zehnder B, Eschenmoser A (1981) Helv Chim Acta 64:1431
135. Wang JJ, Stolowich NJ, Santander PJ, Park JH, Scott AI (1996) Proc Natl Acad Sci USA 93:14320
136. Ichinose K, Kodera M, Leeper FJ, Battersby AR (1993) J Chem Soc, Chem Commun 515
137. Ford SH, Friedmann HC (1976) Arch Biochem Biophys 175:121 and references cited therein
138. Grabau C, Roth JR (1992) J Bacteriol 174:2138 and references cited therein
139. Rémy E, Debussche L, Thibaut D (1996) 4th European Symposium on Vitamin B₁₂ and B₁₂-proteins, Innsbruck, Austria
140. Bernhauer K, Wagner F, Michna H, Rapp P, Vogelmann H (1968) Hoppe-Seyler's Z Physiol Chem 349:1297
141. Kiuchi F, Leeper FJ, Battersby AR (1995) Chem & Biol 2:527
142. Blanche F, Thibaut D, Couder M, Muller J-C (1990) Anal Biochem 189:24